TRAINING CIRCULAR
NO. 90—6—1

C1, TC 90-6-1
HEADQUARTERS
DEPARTMENT OF THE ARMY
Washington, D.C., 30 September 1976

MILITARY MOUNTAINEERING

MILITARY MOUNTAINEERING

INTRODUCTION

This training circular is designed to help the reader become an expert military mountaineer. Training is the single most important aspect in the conduct of successful mountain operations. A commander must never disregard or slight the proper training of his command in mountain operations.

The contents of this training circular are applicable to general, limited and cold war, to include internal defense and development, as well as active nuclear, nonactive nuclear, and nonnuclear warfare. The degree to which the contents of this training circular are applied must be varied to conform to known enemy doctrine, tactics, and actions.

There are mountains in virtually every country in the world and because of this, almost every war that has been fought since the start of recorded history has included some type of mountain operations. It can be assumed that this pattern will not change and troops will again be fighting in mountainous terrain in any future war.

Basic tactics for mountain operations have not changed. What has changed is the equipment and means of transport. The helicopter gives almost immediate access to terrain which in former times was completely out of reach, or could be reached only after hours of tedious climbing. This fact holds true only as long as good weather exists. If the weather turns bad, the commander of military troops must use the old tried and true mountain climbing techniques shown in this circular.

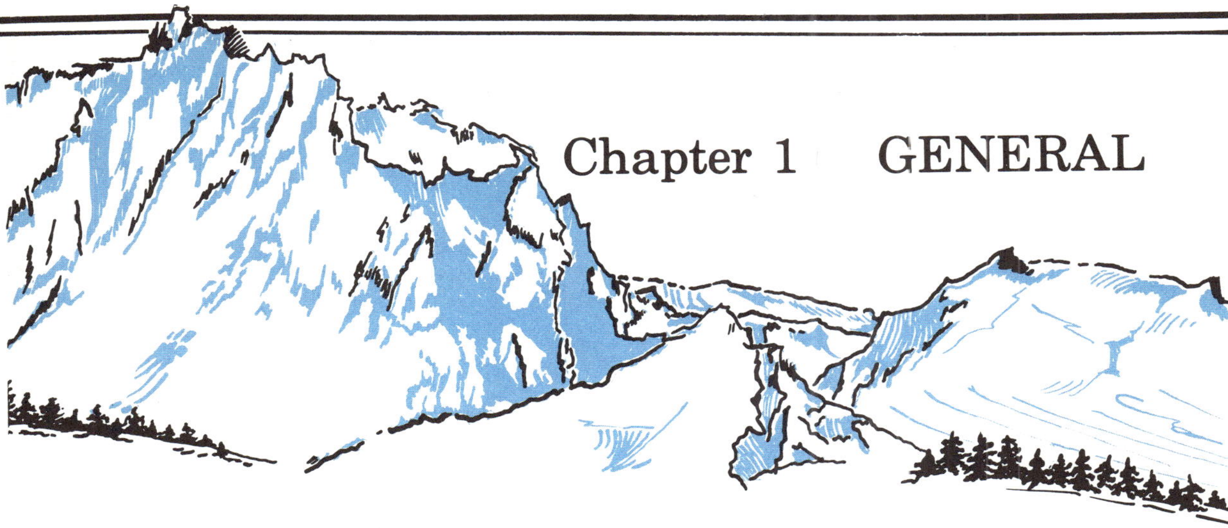

Chapter 1 GENERAL

Section I TERRAIN, CLIMATE AND WEATHER

1—1. Terrain

Mountains are generally defined as land forms higher than 500 meters above the surrounding plain characterized by steep slopes. Slopes are commonly 4 to 30 degrees and in cliffs and precipices, may be vertical or even overhanging. Mountains may consist of an isolated peak, single ridges, or complex ranges extending for long distances like the Rocky Mountains. Most of the terrain is an obstruction to movement. Mountains generally favor the defense. History records many cases in which forces inferior in numbers and equipment have held off superior attackers in mountainous areas.

The following example gives some insight regarding the mountain land masses throughout the world. The land surface of the world covers approximately 91,730,100 square kilometers, of which 4,345,110 are above 3,000 meters. The Central Asian Highlands constitute the largest land mass over 3,000 meters, with close to 1,609,300 square kilometers above this height. South America has some 587,394 square kilometers. All the rest of the land above 3,000 meters is of little consequence on a world-wide basis. However, if the millions and millions of square kilometers of mountains from 500 to 3,000 meters are added, the example becomes exceedingly clear and shows that mountain operations can occur in practically every country in the world in all types of environments.

1—2. Roads and Trails

Existing roads and trails offer the easiest routes in the mountains. Most follow valleys between ridges, crossing ridges and divides by passes. Detailed maps show roads and many of the trails. Terrain intelligence, photo interpretation, local residents, and other sources may supply additional information.

1—3. Cross-Country Movement

Where there are no roads or trails, it is particularly important to have full knowledge of terrain to determine the most feasible routes for cross-country movement. Preoperations intelligence effort must be directed to compilation of topographic or photo map coverage and collection of detailed weather data for the area of operations. When planning mountain operations it may be necessary to use additional information on size, location and characteristics of land forms and drainage, nature and types of rock and soil, and amount and distribution of vegetation.

1—4. Concealment, Cover, and Observation

Irregular mountain topography provides numerous favorable places for concealment and cover. Digging of foxholes and temporary fortifications is generally difficult because soils are thin or stony. Some rock types, however, such as volcanic tuff are readily excavated. In other areas, abundant boulders and other loose rocks can be used for the construction of hasty fortifications. Observation in mountains is variable because of weather and ground cover. Range determination is particularly deceptive.

1—5. Climate

Mountain climate has a very definite effect on the physiology and pathology of the individual because the human body is sensitive to weather changes and differing climates.

1—6. Mountain Air

Mountain air is relatively pure. The higher the elevation, the more nearly pure it becomes. Above 4,500 meters it is practically germ free. The physical composition of the atmospheric air is considerably different at high altitudes from that found at sea level. Forests, especially those with coniferous trees, lessen the percentage of carbon dioxide, and thus do their part in purifying the air. Falling snow also purifies the air by capturing and holding many of the impurities which remain in the air.

High mountain air is dry, especially in the winter when the humidity in the air condenses into ice. This dryness increases with altitude. The amount of water vapor in the air decreases as the altitude increases.

Atmospheric pressure drops as the altitude increases.

The temperature drops and the air becomes more rarefied with altitude.

The sun's rays are either absorbed or reflected by the atmospheric haze which fills the air above low country, especially over cities. The rarefied dry air of the higher altitudes allows all the visible rays of the solar spectrum to pass in their entirety. However, only a fraction of the solar invisible rays can penetrate the atmospheric haze. In pure atmosphere, the proportion of ultraviolet rays remains constant, regardless of the altitude. These conditions increase the possibility of sunburn especially when combined with the existing snow cover.

1—7. Weather

Mountain weather can be either a dangerous obstacle to operations or a valuable aid, depending on how well it is understood and to what extent advantage is taken of its peculiar characteristics. Weather often determines the success or failure of the mission. Mountain weather is highly changeable. Military plans arranged in advance of an operation must be flexible. This is especially true in planning airmobile operations. Every effort must be made to anticipate the weather, and also to allow sufficient latitude in the time schedule so that the leaders of subordinate units can use their initiative in turning the highly important weather factor in their favor. The clouds that frequently cover the tops of mountains and the fogs that cover valleys offer an excellent means of concealing movements which normally would have to be made under cover of darkness or smoke.

The safety or danger of almost all high mountain regions, especially in winter, depends upon a change of a few degrees of temperature above or below the freezing point. Ease and speed of travel are largely dependent on weather. Terrain that can be crossed swiftly and safely one day may become impassable or highly dangerous the next because of snowfall, rainfall, or a rise in temperature. The reverse can happen just as quickly. The prevalence of avalanches depends on terrain, snow conditions, and weather factors.

1—8. Characteristics of Weather

Mountain weather is erratic. Hurricane force winds and gentle breezes may occur just short distances apart. The weather in exposed places contrasts sharply with the weather in sheltered areas. Weather changes in a single day can be so variable that in the same locality one may experience hot sun and cool shade, high winds and calm, gusts of rain or snow, and then perhaps intense sunlight again. This variability results from the life cycle of a local storm or from the movement of traveling storms. In addition, the effects of storms are modified by the following local influences:

- Variation in altitude
- Differences in exposure to the sun and to prevailing winds.
- Distortion of storm movements and the normal winds by irregular mountain topography. These local influences dominate summer storms.

1—9. Temperature

Temperature Inversion. Normally, a temperature fall of from 3°F to 5°F per 300 meter gain in altitude will be encountered. Frequently, on cold, clear, calm mornings when a troop movement or climb is started from a valley, higher temperatures may be encountered as altitude is gained. This reversal of the normal situation is called *temperature inversion.* This condition occurs when air cooled by ice, snow, and heat loss by radiation settles into valleys and low areas. The inversion will continue until sun warms the surface of the earth or a moderate wind causes a mixing of the warm and cold air layers. Temperature inversions are very common in the mountainous regions of the arctic and subarctic.

Solar Heating. At high altitudes, solar heating is responsible for the greatest temperature contrasts. More sunshine and solar heat are received above than below the clouds. The important effect of altitude is that the sun's rays pass through less of the atmosphere and more direct heat is received than at lower levels where solar radiation is absorbed and reflected by dust and water vapor. There may be differences of 40°F to 50°F between the temperature in the sun and that in the shade. Special care must be taken to avoid sunburn and snow blindness which result from the combined action of intense sunlight and the reflected rays from snow fields or clouds. Besides permitting rapid heating, the clear air at high altitudes also favors rapid cooling at night. Consequently, the temperature rises fast after sunrise and drops quickly after sunset. Much of the chilled air drains downward, due to convention currents, so that the differences between day and night temperatures are greater in valleys than on slopes.

1—10. Cloudiness and Precipitation

Cloudiness and precipitation increase with height until a zone of maximum precipitation is reached; above this zone they decrease. Maximum cloudiness and precipitation occur near 1800 meters elevation in middle latitudes and at lower levels as the poles are approached. Usually a dense forest marks the zone of maximum rainfall.

Slopes facing the prevailing wind are cloudier, foggier, and receive heavier precipitation than those protected from the wind, especially when large bodies of water lie to the windward side. However, at night and in winter, valleys are likely to be colder and foggier than higher slopes. Heads of valleys often have more clouds and precipitation than adjacent ridges and the valley floor.

1—11. Wind

In high mountains the ridges and passes are seldom calm; however, strong winds in protected valleys are rare. Normally, wind speed increases with altitude, since the earth's frictional drag is strongest near the ground, and this effect is accentuated by mountainous terrain. Winds are accelerated when they are forced over ridges and peaks or when they converge through mountain passes and canyons. Because of these funneling effects, the wind may blast with great force on an exposed mountainside or summit. In most cases, the local wind direction is controlled by topography.

The force exerted by wind quadruples each time the wind speed doubles; that is, wind blowing at 40 knots pushes four times harder than does a 20 knot wind. With increasing wind strength, gusts become more important and may be 50 percent higher than the average wind speed. When wind strength increases to a hurricane force of 64 knots or more, men should hug the ground during gusts and push ahead during lulls. If a hurricane force wind blows where there is sand or snow, dense clouds fill the air, and rocky debris or chunks of snow crust are hurled along near the surface.

In general, the speed of the winds accompanying local storms is less than that of winds with traveling storms. There are two winds which result from the daily cycle of solar heating. During calm clear days in valleys that are subject to intense solar radiation, the heated air rises and flows gently up the valleys producing a wind called the *valley* or *up-valley* breeze. On clear nights the mountainsides lose heat rapidly and cool the surrounding air which settles downslope to produce the *mountain* or *down-valley* breeze. The down-valley breeze, by pouring cold air into a valley, aids the creation of a temperature inversion.

During the winter season or at extremely high altitudes, commanders must be constantly aware of the *windchill factor* and associated frostbite. Frostbite is the freezing of some part of the body by exposure to temperatures below freezing. It is a constant hazard in operations performed at freezing temperatures, especially when the wind is strong. Usually there is an uncomfortable sensation of coldness followed by numbness. There may be a tingling, stinging, or aching sensation, even a cramping pain. The skin initially turns red. Later it becomes pale gray or waxy white. For all practical purposes, frostbite may be classified as *superficial* or *deep*. Treatment and management are based solely upon this classification. Windchill is covered in detail in FM 31—70 and FM 31—71. During all seasons, exposed areas of the body are subject to windburn or extreme chapping. Although windburn is uncomfortable, it is seldom incapacitating.

1—12. Thunderstorms

Although individual thunderstorms are normally local in nature and usually of short duration, they can be part of a large weather system which may handicap operations in the mountains. In the alpine zone above timberline, thunderstorms may be accompanied by freezing precipitation and sudden squally winds. Ridges and peaks become focal points of concentrated electrical activity which is highly dangerous.

Purely local thunderstorms develop from rising air columns resulting from the intense heating by the sun of a relatively small area. They occur most frequently in the middle or late afternoon. Scattered fair weather clouds of the cumulus type often appear harmless, but when they continue to grow larger and reach a vertical depth of several thousand meters they may turn into thunderstorms on very short notice.

Thunderstorms occurring at night or in the early morning are associated with major changes in the weather situation, which often result in a long period of foul weather before clearing on the high summits. Thunderstorms occurring at these times may also be part of a general storm line and if so, are followed by a prolonged period of cool, dry weather.

1—13. Lightning

Although statistics show that lightning is not one of the major hazards of mountaineering, enough casualties and near casualties have been reported that the matter is at least one of concern. Mountain climbers often find themselves on prominent peaks and exposed ridges, which are particularly subject to lightning strokes and lesser discharges. There are two reasons for this: (1) the ridges help produce the vertical updrifts and the rain-cloud conditions which generate lightning, and (2) the prominences serve to trigger the strokes. There are, however, precautionary measures that can be taken by the climber, and these should be more generally known.

The most obvious way of avoiding lightning in the mountains is not to be on

exposed peaks or ridges, or in an unprotected flat expanse during an electrical storm. If such a storm can be predicted, it is sensible not to climb. Avoid being under prominent or isolated trees. If you are caught in an exposed place, and you have some time before the storm reaches you, you should get as far down the mountain and away from the exposed ridges as you can. Especially avoid those ridges that dominate the skyline. The middle of a ridge is preferable to the end of a ridge.

If lightning strokes seem imminent or are striking nearby, seek at once a place that will protect you from direct strokes and from ground currents. Waste no time. A flat shelf, a slope, or a slightly raised place dominated by a nearby high point would give protection from lightning strokes. If there is any choice, select a spot on dry, clean rock in preference to damp or lichen-covered rock. A scree slope would be very good. Tie yourself in if you are where a severe shock might cause you to fall. The main thing to remember when caught in an electrical storm is to take precautions quickly.

1—14. Fog

On windward slopes, presistant fog, as well as cloudiness and precipitation, frequently continue for days, and are caused primarily by the local barrier effect of the mountain on prevailing winds. Any cloud bank appears as a fog from within. Fog limits visibility and causes whiteout conditions which, in turn, hampers operations by increasing the possibility of accidents. It also facilitates surprise attacks. If fog is accompanied by precipitation, additional clothing will be needed for protection against the uncomfortable combination of cold and wetness. When traveling without landmarks, it will be necessary to use a compass and a map to maintain direction.

1—15. Traveling Storms

The most severe weather conditions, storms involving strong winds and heavy precipitation, are the result of widespread atmospheric disturbances which generally travel in an easterly direction. If a traveling storm is encountered in the alpine zone during winter, all the equipment and skill of the soldier will be pitted against low temperatures, high winds, and blinding snow.

Traveling storms result from the interaction of cold and warm air. The center of the storm is a moving low pressure area where cyclonic winds are generally the strongest. Extending from this storm center is a warm front which marks the advancing thrust of warm air, and the cold front which precedes the onrushing cold and gusty winds. The sequence of weather events, with the approach and passing of a traveling storm, depends on the state of the storm's development, and whether the location of its path is to the north or south of a given mountain area. Generally, scattered cirrus clouds merge into a continuous sheet which thickens and lowers gradually until it becomes altostratus. At high levels, this cloud layer appears to settle. Lower down, a stratus deck may form overhead. A storm passing to the north may bring warm temperatures with southerly winds and partial clearing for a while before colder air with thundershowers or squally conditions moves in from the northwest. However, local cloudiness often obscures frontal passages in the mountains. The storm may go so far to the north that only the cold front phenomena of heavy clouds, squalls, thundershowers, and colder weather are experienced. The same storm passing to the south would be accompanied by a gradual wind shift from northeasterly to northwesterly, with a steady temperature fall and continuous precipitation. After colder weather moves in, the clearing at high altitude is usually slower than the onset of cloudiness, and storm conditions may last several days longer than in the lowlands.

Rapidly changing weather conditions often create glaze, a coating of ice which forms on exposed objects. Glaze occurs under special storm conditions when light rain or drizzle falls through air below 32°F., and strikes a surface that also is below 32°F., freezing to the surface in the form of glaze. Glaze usually forms near the warm front of a storm and only persists if colder weather follows.

1—16. Weather Forecasting

The use of the portable aneroid barometer, thermometer, and hygrometer can be of great assistance in making local weather forecasts. Reports from other localities and from any weather service including Air Force, Navy, or Weather Bureau are also of great value.

In order to be utilized to its fullest extent, the resulting forecast must get down to the small unit leaders who are expected to make use of weather favorable for specific missions.

1—17. Bad Weather

Most of the bad weather experienced in mountain regions is a result of—

- Local storms in the form of thunderstorms with or without showers.
- Traveling storms which may be accompanied by radical and severe weather changes over a broad area. Usually, each type of storm may be identified by the clouds associated with it.
- Seasonal moisture-bearing winds of the monsoon type which bring consistently bad weather to some mountain ranges for weeks at a time.

1—18. Cloud Types

It is possible to differentiate between local and traveling storms, and estimate their probable occurrence, by recognition of the various cloud types (fig. 1—1). The heights shown in the figure are approximate mean heights and may vary based on geographical location.

Cirrus clouds are composed of ice crystals and may occur from 6,500 to 12,000 meters high. They may be detached white clouds with fibrous structure or they may be extensive thin veils.

Altostratus clouds are dark clouds that form a continuous uniform sheet at elevations of 2,500 to 6,000 meters. They may cover the sky completely and develop from the descending and thickening cirrus or the merging of high cumulus types.

Cumulus clouds resemble great patches of cotton or giant heads of cauliflower and form at any height from 500 to 4,000 meters. Their bases are flat, but their tops vary in size and height. Cumulus clouds are brilliant white in direct sunlight, but dark on the shaded side. Altocumulus clouds are similar in form, but smaller, and appear in uniform layer or cloud deck at 2,500 to 7,000 meters.

Cumulonimbus (thunderheads) are overgrown and darkened cumulus clouds that are yielding, or are likely to yield precipitation. A single cumulonimbus cloud mass may extend from a base at 800 meters elevation to a height of 10,000 or 11,000 meters. The tops of cumulonimbus clouds are made up of ice crystals and have a fibrous texture.

Stratus clouds are similar to altostratus, but much lower. They may develop from a fog layer in which the bottom portion has evaporated, or where upslope winds are blowing. Often they are associated with moisture precipitation and usually are seen below 900 meters except where they are forced up over a mountain or lie in high valleys.

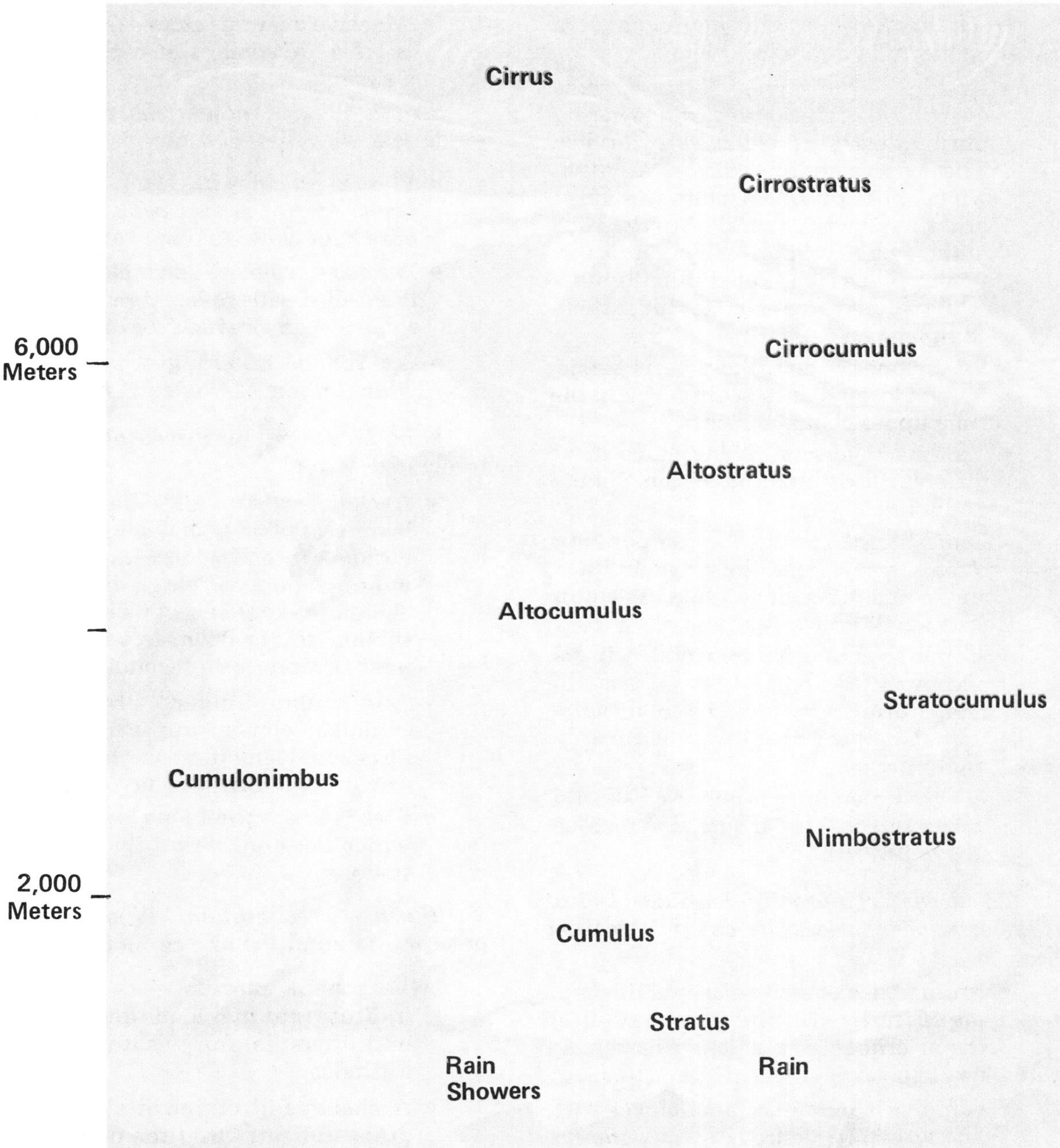

Figure 1—1 *Cloud Types*

1—19. Weather Predictions

Each of the cloud types characterizes one or several phases in a given weather situation. The following are the most common weather indications, although they are not always applicable. An individual remaining in one mountain region for several weeks in any season can add indications for that area based on his own experience with local weather changes. As an example, showery type precipitation is often produced by cumulus or cumulonimbus clouds and frequently produces sudden wind gusts and large temperature fluctuations.

Traveling Storms. The approach of a traveling storm is indicated when :

- A thin veil of cirrus clouds spreads over the sky, thickening and lowering until altostratus clouds are formed. The same trend is shown at night when a halo forms around the moon and then darkens until only the glow of the moon is visible. When there is no moon, cirrus clouds only dim the stars while altostratus clouds hide them completely.
- Low clouds which have been persistent on lower slopes begin to rise at the time upper clouds appear.
- Various layers of clouds move in at different heights and become more abundant.
- Lens-shaped clouds accompanying strong winds lose their streamlined shape and other cloud types appear in increasing amounts.
- A change in the direction of the wind is accompanied by a rapid rise in temperature not caused by solar radiation. This may also indicate a warm, damp period.
- A light green haze is observed shortly after sunrise in mountain regions above timberline.

Local Disturbances. Indications of local thunderstorm showers, or squally weather are :

- An increase in size and rapid thickening of scattered cumulus clouds during the afternoon.
- The approach of a line of large cumulus or cumulonimbus clouds with an advance guard of altocumulus clouds. At night, increasing lightning windward of the prevailing wind gives the same warning.
- Massive cumulus clouds hanging over a ridge or summit at night or in the daytime.

Strong Winds. Indications from strong winds seen at a distance may be :

- Plumes of blowing snow from the crests of ridges and peaks or ragged shreds of cloud moving rapidly.
- Persistent lens-shaped clouds, or a band of clouds, over high peaks and ridges or downwind from them.
- A turbulent and ragged banner cloud which hangs to the lee of a peak.

Fair Weather. Fair weather may be associated with :

- A cloudless sky and shallow fog or layers of smoke or haze at valley bottoms in early morning; or from a vantage point of high elevation, a cloudless sky that is quite blue down to the horizon or down to where a level haze layer forms a secondary horizon.
- Conditions under which small cumulus clouds appearing in the forenoon do not increase, but decrease or vanish during the day.
- Clear skies, except for a low cloud deck which does not rise or thicken during the day.

During Precipitation. When there is precipitation and the sky cannot be seen :

- Very small snowflakes or ice crystals indicate that the clouds above are thin and there is fair weather at high altitudes.
- A steady fall of snowflakes or raindrops indicates that the precipitation has begun at high levels and that bad weather is likely to be encountered on ridges and peaks.

Section II ACCLIMATIZATION AND CONDITIONING

1—20. Necessity for Acclimatization and Conditioning

General. In general, the training of soldiers in mountains of low or medium elevation (1500 to 2500 meters) does not require the special conditioning and acclimatization procedures discussed here. However, some individuals because of poor physical condition or other reasons will find that they will have some impairment of operating efficiency at these low altitudes. Above 2500 meters practically all unacclimatized soldiers may be expected to display some altitude effects.

Acclimatization. A person is said to be *acclimatized* to high elevations when he can live and *perform physically* and *psychologically* at the elevation in *reasonable comfort* and *effectiveness.* The acclimatization process begins immediately upon arrival at the higher elevation. If the change in elevation is large and abrupt, a proportion of the exposed soldiers will suffer the symptoms of acute mountain sickness. Disappearance of the symptoms (4—7 days) does not indicate complete acclimatization. The process of adjustment continues for weeks or months. The limiting altitude to which complete acclimatization is possible is approximately 5,500 meters.

Work Capacity. Immediately upon arrival at high elevations only moderate physical work can be performed because of extreme breathlessness. *Therefore, the expectation that freshly deployed, unacclimatized troops can go immediately into action is at best unlikely, and at worst could end in disaster if such a unit were to fight against an acclimatized enemy.* Even after several weeks of living at 4,000 meters, the maximum rate of work will only be 70% of the rate of work at sea level, and approaches sea level values only after months or years of continuous stay. The deficit is, of course, greater at higher elevations.

Behavioral Effects. Listed below are some of the behavioral effects that will be encountered in unacclimatized personnel:

- Increased errors in performing simple mental arithmetic.
- Decreased ability for sustained concentration.
- Deterioration of memory.
- Decreased vigilance.
- Increased irritability in some individuals.
- Impairment of night vision and some constriction in peripheral vision.

Self-evaluation is impaired in the same manner as if the person was intoxicated. During the first few days at high altitude, leaders will have extreme difficulty in maintaining a coordinated, operational unit. The roughness of the terrain and the harshness and variability of the weather add greatly to the difficulties of unacclimatized personnel. Although strong motivation may succeed in overcoming some of the physical handicaps imposed by the environment, the total impact will still result in errors of judgement by individual soldiers and officers alike.

Deployment to High Elevations. Since there is no way to shorten the acclimatization process, and since the absence of acclimatization hampers the successful execution of operations, the following possibilities are practicable:

- Movement upward by stages, spending 2 or 3 days at each stage; the first stage beginning at the 2,500 meter level, and subsequent stages being at each succeeding 600 to 900 meter increment until the ultimate destination is reached. By this time, a considerable degree of acclimatization will have occurred, mountain sickness greatly reduced, and the operational potential greatly increased.

- When available from medical personnel, pretreatment with Carbonic Anhydrase inhibitors such as acetazolamide 250 mg twice or four times a day will reduce incidence and severity of acute mountain sickness from 40% to 90% depending upon the height of deployment and the number of days of pretreatment.
- Initial acclimatization at a high altitude in CONUS or another safe area and deployment to the operational site from this area by rapid transport.
- Troops might be moved directly to high altitudes if allowances can be made for relative inactivity for the first 3 to 5 days prior to commitment.

Psychological adjustment. Many persons who have lived in lower altitudes all their lives may have preconceived notions about the supposedly harmful effects of high altitudes on a human. To them any illness, however trivial, may be construed as an ill effect of altitude which may cause them to become unduly concerned about their physical condition. This can be prevented by educational programs of high altitude effects, which are usually not harmful and of short duration. Many men who have lived on flat terrain have difficulty when faced with the problem of learning to negotiate steep slopes or cliffs. They will develop a sense of insecurity and fear. These men must be gradually introduced to these new terrain features and encouraged to develop their confidence until they can negotiate these obstacles with assurance and ease. Men must be taught the various climbing techniques and be indoctrinated with the principles of mountain marching. There are many individuals who possess fear of height in varying degrees. Many of these individuals will overcome this fear by familiarizing themselves with the problem. *The individual cannot be forced to disregard this fear.*

Physical conditioning. Regardless of previous Army training and the amount of flat cross-country marching practice, the soldier newly initiated into the mountains finds mountain marching hard and tiring. New muscles are brought into use and must be developed and hardened. In addition, a new technique of rhythmic movement must be learned. This conditioning is attained only through frequent marches and climbs, with personnel carrying their TOE and special equipment loads. This will condition the back as well as the legs and result in increased capability and endurance. Simultaneously with this development, men acquire increasing confidence and better ability to safely negotiate terrain which they had previously considered impassable.

1—21. Personal Hygiene

The principles of personal hygiene and sanitation that govern the operations of troops in more normal terrain are also applicable in the mountains. Commanders must conduct frequent inspections to insure that personal habits of hygiene are not being disregarded.

- *Personal cleanliness* is especially important in the high mountains and particularly during periods of cold weather. In freezing temperatures, the individual has a tendency to neglect washing due to both the cold and the scarcity of water. This can result in skin infections and vermin infestation. If bathing is impossible for any extended period of time, the soldier should, as a minimum, examine his skin and stimulate and cleanse it as much as possible. This practice will help reduce the possibility of skin infection and will also make the individual feel better in general.
- The principles of *foot hygiene* apply with greater force in the mountains than they do in flat terrain. Particular attention must be taken to protect the feet from cold injuries. The causative factors for these injuries are present throughout the entire year in high mountains. Feet should be kept dry and socks changed as often as necessary. Generally, boots should be laced tighter than normal for climbing to provide necessary support. Care

must be taken to ensure that there are no wrinkles in the socks as this will quickly cause blisters.

- During periods of extreme cold, there is a general tendency for the individual to permit himself to become *constipated.* This condition is brought about by the desire to avoid the inconvenience and discomfort of relieving himself. Adequate water intake plus a low protein-high roughage diet may be helpful in preventing constipation.

1—22. Water Supply

The common opinion that mountain water is safe for consumption is not always true. This dictates the need for additional emphasis on training in water discipline to insure that the soldier drinks water only from approved sources. Fluids lost through respiration, perspiration, and urination must be replaced if the soldier is to operate with any efficiency. Maintenance of fluid balance is a recognized problem. The sense of thirst may be blunted by high elevations despite the greater threat of dehydration. Hyperventilation and the cool, dry atmosphere bring about a three to four fold increase in water loss by the lungs. Hard work and overheating increase the perspiration rate. A conscientious effort must be made to drink fluids even in the absence of a sensation of thirst. It is recommended that one quart of water, or equivalent, be drunk every four hours. Emphasis must be placed on each of the three rules of water discipline.

- Drink only treated water.
- Conserve water for drinking. Potable water in mountains may be in short supply.
- Do not contaminate or pollute water sources.

1—23. Nutritional Acclimatization

The following elements are characteristic to nutritional acclimatization in mountain operations:

- Weight loss during the first few days at high elevations is characteristic.
- A loss of appetite is characteristically associated with symptoms of mountain sickness.
- Loss of weight generally stops with acclimatization.
- Higher carbohydrate diets are indicated.
- At progressively higher elevations, the tolerance of fatty foods decreases rapidly.

1—24. Mountain Sickness

Mountain sickness is an acute temporary illness which *affects* the *novice* and *experienced* climber alike. Personnel are subject to this sickness in altitudes as low as 2,500 meters. Incidence and severity increases with altitude and when transport to high altitudes is rapid. Indications are that substantial disability and ineffectiveness can occur in *50 to 80 percent* of troops rapidly brought to altitudes in excess of 4,000 meters. At lower altitudes or where ascent to altitudes is gradual, the lessened severity of the sickness will permit a majority to carry out assignments with moderate effectiveness, although with some discomfort.

The new arrival at moderately high elevations (3000 to 4000 meters) commonly feels well for the first few hours; a feeling of exhileration or well-being is not unusual. There may be an initial awareness of breathlessness upon exertion and a need for frequent pauses to rest. Breathing irregularities can occur, particularly during sleep. Individuals who become aware of these changes may exhibit some apprehension. The true onset of symptoms begins 4 to 12 hours after arrival at the higher altitude, with feelings of :

- sickness
- sluggishness
- headache

Headache is the most distressing symptom and may be quite severe. Even when headache is not present there is almost always some loss of appetite and decrease in tolerance for food. Nausea, even without food intake, occurs and leads to a lessening of food intake. Vomiting may occur and will contribute to a tendency to become dehydrated. Despite fatigue, a person will be unable to sleep. The symptoms usually develop and increase to a peak intensity by the second day and then gradually subside over the next several days so that the total course may extend to 5 to 7 days. In some instances the headache will persist until the patient is returned to a lower level.

Treatment consists of :

- Relieving the headache with aspirin or an equivalent. The patient should wear sunglasses since bright sunlight seems to aggravate the headache.
- Reassuring the patient.
- Encourage fluids and light foods in small frequent amounts.
- Symptoms may be alleviated by moving the patient to lower altitudes (600 to 900 meters). This also provides for a more gradual acclimatization.

1—25. High Altitude Pulmonary Edema (HAPE)

HAPE occurs solely under conditions of exposure to low partial pressures of oxygen, has been encountered exclusively at high elevations; and characteristically occurs in individuals experiencing excellent health. HAPE may be considered as a form or manifestation of acute mountain sickness since it occurs during the period of susceptibility to this disorder.

- HAPE as currently recognized, is characterized by *high mortality.*

Incidence and severity both increase with altitude. Except for acclimatization to altitude there are no known factors to indicate spontaneous resistance or immunity. Cases are rare on the day of arrival and infrequent after the fourth day. Very few cases have been reported after more than ten days at high altitude. The incidence of HAPE is much less frequent than that of mountain sickness.

Factors contributing to HAPE include :

- A history of prior high altitude pulmonary edema.
- A rapid or abrupt transition to high altitude.
- Strenuous physical exertion.
- Exposure to cold.
- Anxiety.

HAPE symptoms

- Progressive dry cough which becomes productive with frothy white or pink sputum.
- Cyanosis, or the turning blue of the face, hands, and feet will occur and will become more intense.
- Majority of victims have the symptoms of acute mountain sickness, therefore, the early pulmonary difficulties may not be primary or striking.
- Personnel with acute mountain sickness should be closely observed for increasing *ill feeling, labored breathing,* repeated *clearing* of the throat, development of the *dry cough* and an increase of *difficult* or *labored breathing* at night.
- In rapidly progressive cases the onset of respiratory difficulty may be quite sudden, associated with a choking feeling and rapid deterioration.

HAPE treatment

- Prompt treatment consisting of *rest, warmth,* and the *administration of oxygen.*
- Immediate evacuation to *lower altitudes.*
- Depending upon the ease versus hazards of evacuation, the use of rest, warmth, and oxygen may be preferable to a hasty evacuation to lower elevations.
- If *untreated,* HAPE may become *irreversible* and lead to a *fatal outcome.* Cases that are recognized early and treated promptly may be expected to recover. HAPE can be largely prevented by a program of acclimatization, as indicated under acute mountain sickness. Lastly, individuals who have had a previous attack of HAPE are more prone to have second attacks.

Chapter 2

MILITARY MOUNTAINEERING

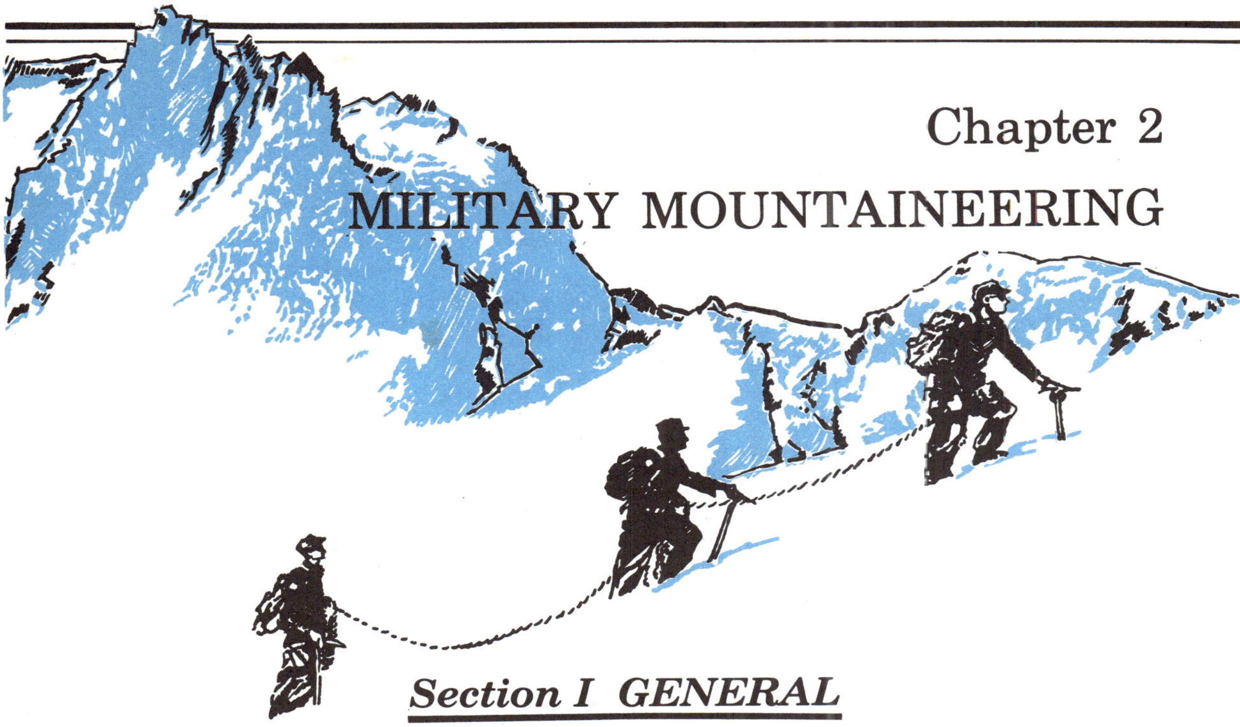

Section I GENERAL

2—1. Importance

Military mountaineering provides access to otherwise inaccessable rugged mountainous terrain. In such terrain, there are three adversaries :

- the enemy
- the mountains
- the weather

These adversaries can, however, be overcome by troops with specialized training who are skilled in the use of special mountain climbing equipment and techniques.

2—2. How It Is Used

Soldiers skilled in military mountaineering can perform many essential missions in difficult terrain. They can serve as guides, observers, snipers, patrols, security elements, and assault team members.

2—3. Command Understanding

An estimate of friendly or enemy capabilities in rugged terrain is possible only if the higher commanders are thoroughly familiar with the capabilities of the military mountaineer. All leaders who are to operate in mountainous areas must be oriented in this respect. All leaders at company level should become skilled military mountaineers.

2—4. Training Requirements

Specialists

Each platoon size unit should have at least two teams of three expert rock climbers (assault climbers) while the remainder of the unit should be skilled in at least basic military mountaineering (mountain walking, balance climbing, knot tying, construction of installations and bridging).

Semi-trained Climbers

Preparation for mountain operations is not complete until the unit commander knows how all his men will react to exposure to height and to the out-of-the-ordinary muscular effort. A few hours of climbing for all men provides a valuable index to each man's future reactions. Each man is exposed to the unknown, and therefore frightening dangers. Fear of falling is a basic instinct, and may be as hard to overcome as fear of enemy fire. Analysis of a man's reaction to height will enable the commander to place those who do not overcome the fear of falling in a position where they will not jeopardize the safety of the unit. *No individual can be forced to disregard his fear of height.*

Section II BASIC TECHNIQUES

2—5. Mountain Walking

Mountain walking can be divided into four different techniques, dependent on the general formation of the terrain. These techniques are walking on hard ground, grassy slopes, scree slopes, and talus slopes. Included in all of these techniques are certain fundamentals which must be mastered in order to obtain a minimum expenditure of energy and loss of time. These fundamentals are—

- the weight of the body must be kept directly over the feet.
- the sole of the boot must be placed flat on the ground.

This is most easily accomplished by taking small steps at a slow steady pace. An angle of descent which is too steep must be avoided and small indentations in the ground, however small, should be used to advantage.

Walking on Hard Ground. Hard ground is generally considered to be firmly packed dirt which will not give away under the weight of a man's step. When ascending, the above fundamentals should be applied with the following addition: *the knees must be locked on every step in order to rest the muscles of the legs.* When steep slopes are encountered, they can be traversed rather than climbed straight up. Turning at the end of each traverse should be done by stepping off in the new direction with the uphill foot. This prevents crossing the feet and possible loss of balance. In traversing, the full sole principle is most easily accomplished by rolling the ankle away from the hill on each step. For narrow stretches the herringbone step may be used; that is, ascending straight up a slope with toes pointed out and using all other principles mentioned thus far. Descending is most easily done by coming straight down a slope without traversing. The back must be kept straight and the knees bent in such a manner that they take up the shock of each step. Again, it must be remembered that the weight must be directly over the feet, and that the full sole must be placed on the ground with every step. Walking with a slight forward lean and with feet in a normal position will make the descent easier.

Grassy Slopes. Grassy slopes will usually be made up of small tussocks of growth rather than one continuous field. In ascending, it will be found that while the techniques previously mentioned are applicable, it is better to step on the upper side of each hummock or tussock where the ground is more level than on the lower side. Descending is best accomplished by traversing. At times a hop-skip may be also used to advantage. The hop-skip is a hopping movement in which the lower foot takes all the weight and the upper foot is used for balance only. The hop-skip is also useful on hard ground and on scree when descending.

Scree Slopes. Scree slopes consist of small rocks and gravel which have collected below rock ridges and cliffs. The size of the scree varies from sand to pieces the size of a fist. Occasionally it occurs in mixtures of all sizes but normally scree slopes will be made up of the same size particles. Ascending scree slopes is difficult, most tiring, and should be avoided whenever possible. All principles of ascending hard ground apply, but each step must be picked carefully so that the foot will not slide down when weight is placed on it. This is best accomplished by kicking in with the toe of the upper foot so that a step is formed in the scree. After determining that the step is stable, carefully transfer weight from the lower to the upper foot and repeat the process. The best method for descending scree is to come straight down the slope using a short shuffling step with the knees bent, back straight, and feet pointed downhill. When several climbers descend a scree slope together they should be as close together as possible, one behind the other, to prevent injury from dislodged rock. Since there is a tendency to run down scree, care must be taken that this is avoided and control is not lost. When a scree slope must be traversed and no gain or loss of altitude is desired, the hop-skip method should be used.

Talus Slopes. Talus slopes are similar in makeup to the scree slopes, except that the rock pieces are larger. The technique of walking on talus is to step on top of, and on the uphill side of, the rocks. This prevents them from tilting and rolling downhill. All other fundamentals previously mentioned are applicable.

General Precautions. It is of the utmost importance that rocks are not kicked loose. Falling rocks are extremely dangerous to the men below and also make a great deal of noise. If during a climb a rock is kicked loose it is important that the warning word "ROCK" be given to allow those below an opportunity to move out of the path of the rolling rock. Often, however, there is not enough time for the climber to move out of the path of the falling rock. If this is the case, the climber should lean into the slope to minimize his exposure to the falling rock. Climbers must be aware of the danger of looking up upon hearing the warning "ROCK", because this can result in facial and head injuries. Carelessness can cause the failure of a well planned mission, since one rock no bigger than a man's head can kill or severely injure several men as well as ruin all security measures. Stepping over rather than on top of obstacles, such as large rocks and fallen logs, will help much toward avoiding fatigue. Usually it will be found that talus is easier to ascend and traverse, while scree is a more desirable avenue of descent.

RIGHT

WRONG

Figure 2—1 Right and wrong balance climbing body position.

398-868 O - 83 - 2

Figure 2—2 **Descent.**

2—6. Balance Climbing

General. Balance climbing is the type of movement used to climb rock faces. It is a combination of the balanced movement of a tight-rope walker and the unbalanced climbing of a man ascending a tree or ladder. In balance and party climbing, the climber must study the route he is to travel in order to make sure he has selected the best route and has the necessary equipment. During the process of route selection, he should mentally climb the route so he will know what to expect. Climbers *should not* wear gloves when balance climbing. They *should,* however, wear gloves for all type rappels to protect the palms from rope burns.

Body Position. (fig. 2—1). The soldier must climb with the body in balance, which means that the weight should be placed over the feet as he moves. The feet, not the hands should carry the weight except on the steepest cliffs. The hands are for balance. Feet will not hold well when the climber leans in toward the rock. With the body in balance, the climber moves with a slow, rhythmic motion. Three points of support such as two feet and one hand, are used whenever possible. Hand holds that are waist to shoulder high are preferable. Relaxation is necessary because tense muscles tire quickly; when resting, the arms are kept low where circulation is not impaired. Use of small intermediate holds is preferable to stretching and clinging to widely separated big holds. A spread-eagle position, in which a man stretches so far he cannot let go, should be avoided.

Descents. (fig. 2—2). In descents, the climber faces out where the going is easy, sideways where it is hard, and faces in where it is difficult. He uses the lowest possible handholds.

Types of Holds.

- *Push holds.* (fig. 2—3). Push holds are pushed down upon and help the climber keep his arms desirably low, but are more difficult to hold onto in case of a slip. A push hold is often used to advantage in combination with a pull hold.
- *Pull holds.* (fig. 2—4). Pull holds are those that are pulled down upon and are the easiest holds to use. They are also the most likely to break out.
- *Friction holds.* (fig. 2—5). Friction holds are those dependent solely on the friction of hands or feet against a

relatively smooth surface with a very shallow hold. They are difficult to use because they give a feeling of insecurity which the inexperienced climber tries to correct by leaning close to the rock, thereby increasing his insecurity. They often serve well as intermediate holds, some of which will give needed support while the climber moves over them, but would not hold him were he to stop.

- *Jam holds.* (fig. 2—6). Jam holds involve jamming any part of the body or extremity into a crack. This can be done by putting the hand into the crack and clenching it into a fist or by putting the arm into the crack and twisting the elbow against one side and the hand against the other side. When using the foot in a jam hold care must be taken to insure that the boot is not jammed into the crack in such a way the it cannot be removed easily when climbing is continued.
- *Combination holds.* The holds previously mentioned are considered basic and from these any number of combinations and variations can be used. The number of these variations depends only on the limit of the individual's imagination. Here are a few of the more common ones:

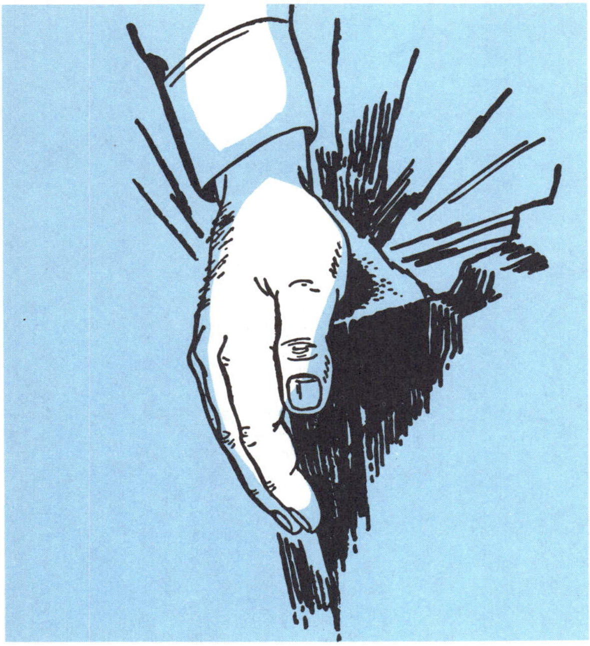

Figure 2—3 Push hold.

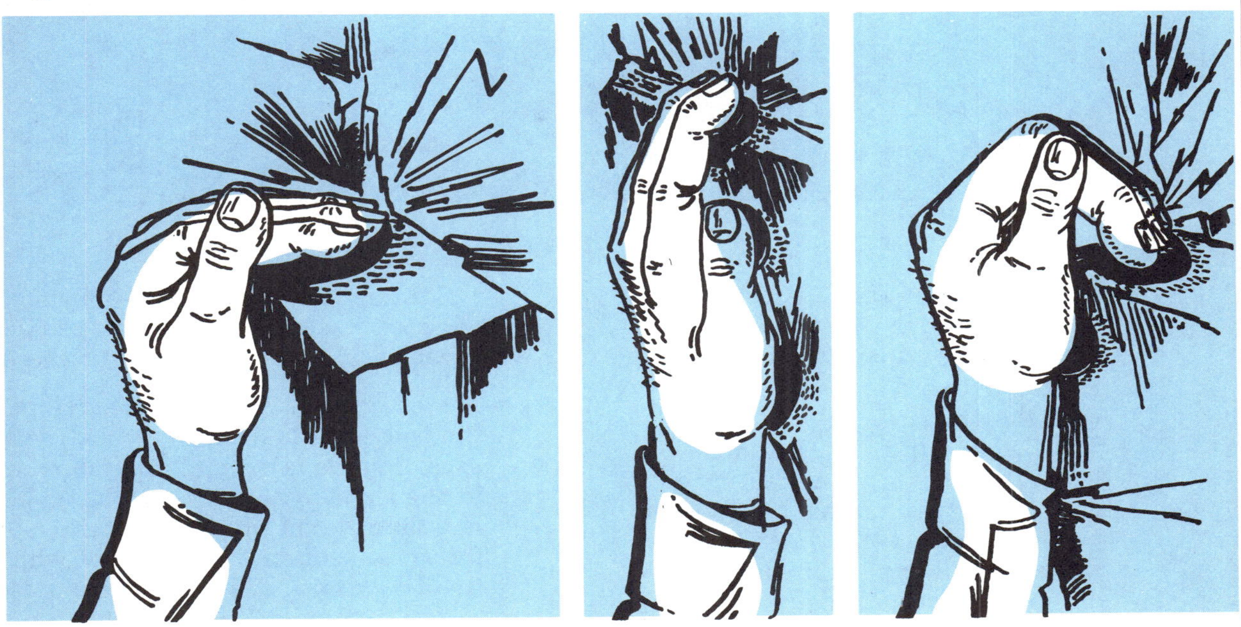

Figure 2—4 Pull hold.

Figure 2—5 ***Friction hold.***

- The pinch hold (fig. 2—7) attained by pinching a protruding part between the thumb and fingers. Pulling outward or pressing inward with the arms.
- The lie-back (fig. 2—8) is done by leaning to one side of an offset crack with the hands pulling and the feet pushing against the offset side.
- Inverted pull or push holds (fig. 2—8) sometimes called underholds, permitting cross pressure between hands and feet.
- Chimney climbing (fig. 2—9) where cross pressure is exerted between the back and the feet or hands or knees.
- *Footholds.* (fig. 2—10). On steep slopes the body should be kept vertical, with use being made of small irregularities in the slope to aid friction. Footholds less than 1½ cm wide can be sufficient for intermediate holds, even when they slope out.
- *Shoulder stand.* (fig. 2—11). The shoulder stand or human ladder, is

used to overcome a holdless lower section of a pitch in order to reach the easier climbing above. The lower man is anchored to the rock and belays the leader who uses his body as a ladder to overcome the difficult pitch.

- *Movement on slabs.* Slab is a relatively smooth portion of rock laying at an angle. When traversing, the lower foot is pointed slightly downhill to increase balance and friction of the foot. All irregularities in the slope should be utilized for additional friction. On steep slabs it may be necessary to squat with the body weight well over the feet and hands used alongside for added friction. This position may be used for ascending, traversing, or descending. A slip will result if you lean back or let the buttocks drag. Wet, icy, mossy, or scree covered slab is most dangerous.

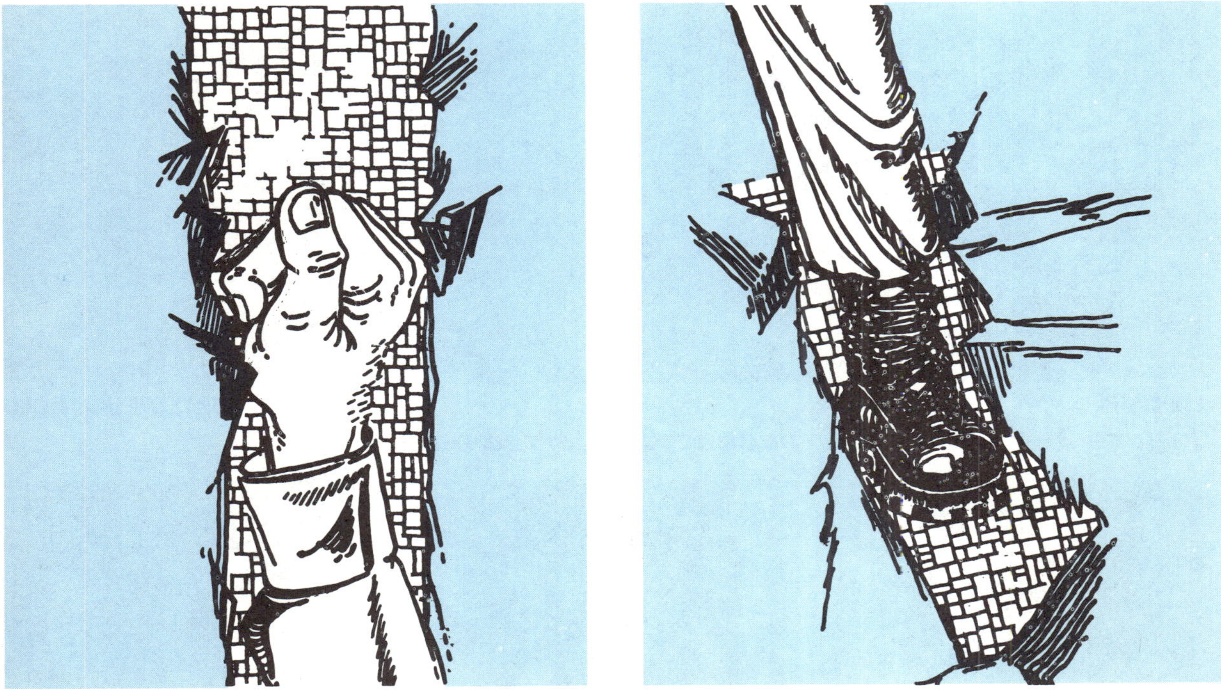

Figure 2—6 ***Jam hold.***

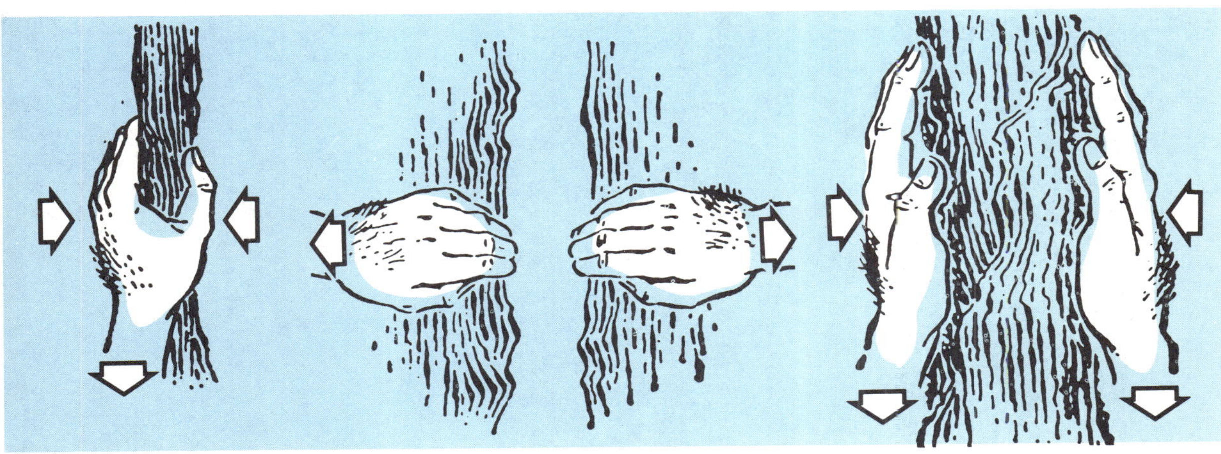

Figure 2—7 ***Pinch hold and cross pressure between hands.***

LIE-BACK

INVERTED PULL HOLD

Figure 2—8 *Lie-back and inverted pull hold.*

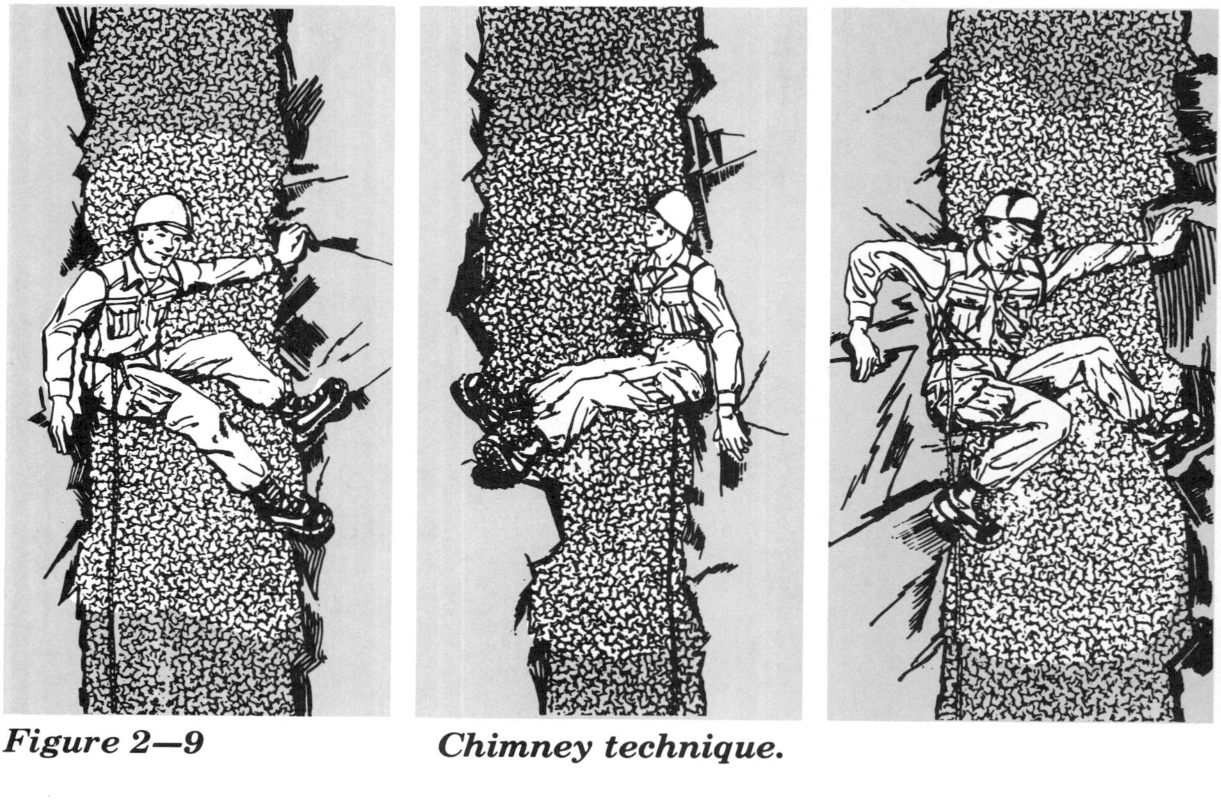

Figure 2—9 *Chimney technique.*

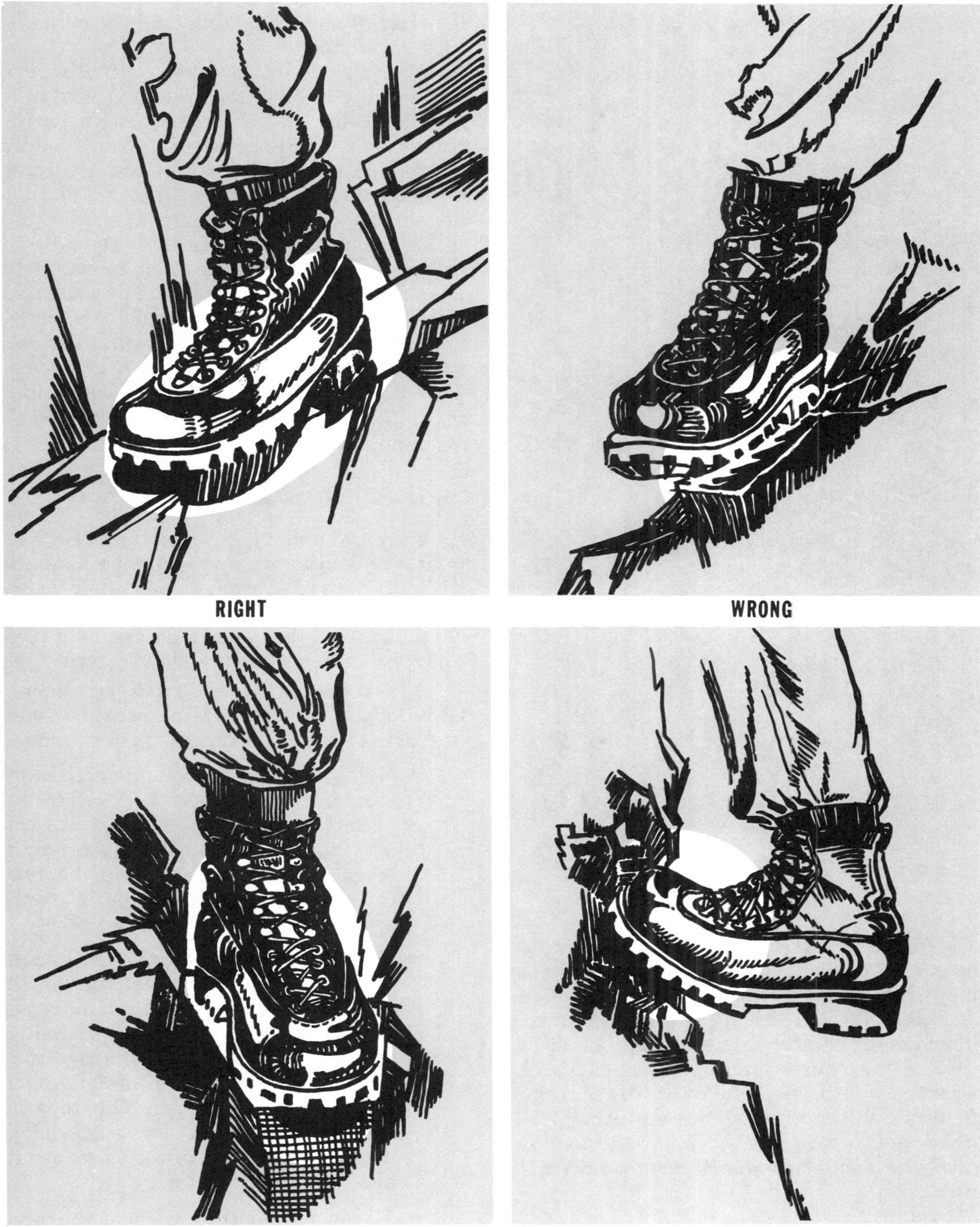

Figure 2—10 *Footholds.*

Figure 2—11 Shoulder stand.

Use of Holds. A hold need not be large to be good. The climber must not try to skip or jump from one to another. It is, however, often desirable while traversing to use the hop step, in which the climber changes feet on a hold so that he may move sideways more easily. A slight upward hop followed by precise footwork will accomplish this useful step. *All hand holds and foot holds are tested by applying weight gradually before they are used.*

Margin of Safety.

A margin of safety is the protective buffer a climber keeps between what he knows to be the limit of his ability and what he actually tries to climb.

The climber learns his margin of safety by climbing close to the ground, or tied to a rope and belayed by a trained man above. He climbs first on the easy and obvious holds, next on the more difficult ones, and finally on difficult pitches until he reaches the limit of his ability.

The margin of safety should be calculated not only for the pitch immediately ahead but for the entire climb. The route and movements of the climber should be planned enough in advance so he never finds himself in situations beyond his ability to overcome. *The leader* of a group *should know the capabilities* of his men and *make allowances* for their limitations.

2—7. Climbing Ropes and Knots

Purpose. Much of the climbing in military operations may be balance climbing without a rope. However, on some steep unbroken cliffs where exposure is great, climbing with the rope is necessary. Fixed ropes and other aids may also be needed.

Types and Characteristics of Ropes. Units engaged in mountain operations will find use for several different types of rope.

- Nylon rope is most commonly used in climbing. The rope is 1.1 cm in diameter and is issued in 36½ meter lengths. The actual breaking strength when dry averages 3,840 pounds (± 5%). The breaking strength is reduced by 18% when the rope is wet.
- Nylon sling rope is commonly used in 4 meter lengths. It is the same diameter as the nylon climbing rope.

  Manila rope 1.9 cm in diameter and larger will be used in the construction of various types of installations requiring great length. The rope is better than nylon for suspension traverses and rope bridges because it has less stretch factor.
- All ropes when they are bent sharply, as around a snaplink, lose a portion of their strength at the bend.

Terms Used in Rope Work (fig. 2—12)

A *bight* of rope is a simple bend of rope in which the rope does not cross itself.

A *loop* is a bend of rope in which the rope does cross itself.

A *half hitch* is a loop which runs around an object in such a manner as to lock itself.

The *running end* of the rope is the free end of the rope.

The *standing part* of the rope is the static part.

The *lay* of the rope is the same as the twist of the rope.

Care of Rope. Because the rope frequently is the climber's lifeline, it deserves a great deal of care and respect.

The rope should not be stepped on or dragged on the ground. Small particles of dirt will be ground between the strands and will slowly cut them.

The rope should not be in contact with sharp corners or edges of rock which may cut it.

Keep the rope dry as much as possible. If it has become wet, dry it as soon as possible.

Do not leave the rope knotted or tightly stretched longer than necessary, and do not hang it on sharp edges, such as nails.

When using ropes in installations (suspension traverses, rope bridges, etc.), do not let them rub against one another, as this may cut and burn them. Whenever such installation ropes are wet, tension should be relieved.

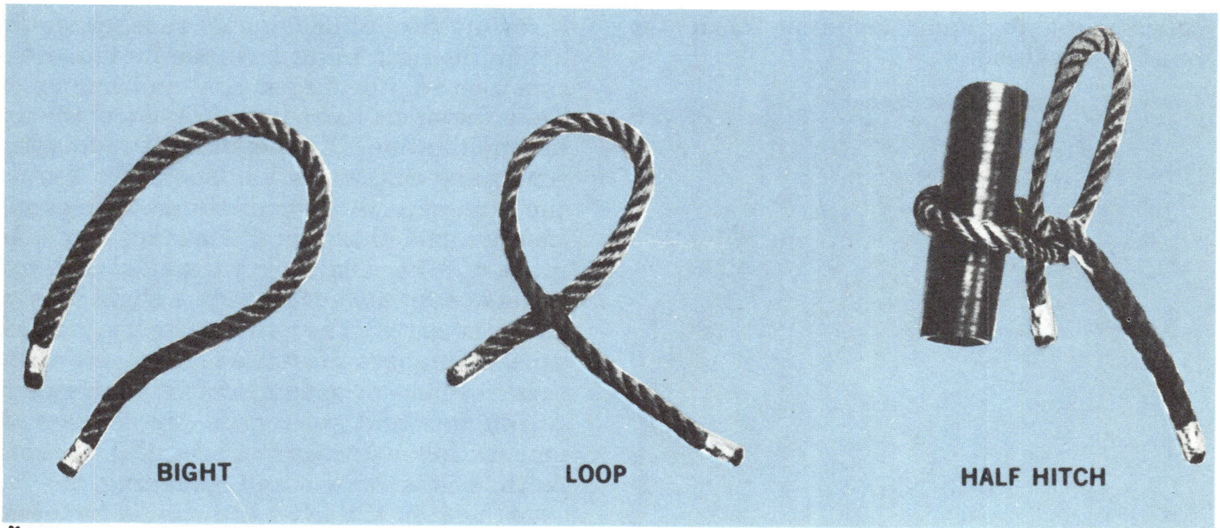

Figure 2—12

Rope Management (fig. 2—13). Before using a rope it should be inspected for frayed or cut spots. If such a spot is found, the rope should be whipped on both sides of the bad spot and then cut. Climbing rope should never be spliced, since it will be hard to manage at the spliced point. Rope should always be coiled except when in actual use. New climbing rope should be marked in the middle by the same process used for whipping the ends. New sling rope as well as any other rope that has been cut from a long piece should be whipped at the ends. New rope is often stiff and should be worked with in order to loosen it up. By placing a mark at the middle of the rope, the exact center of the rope can quickly be determined.

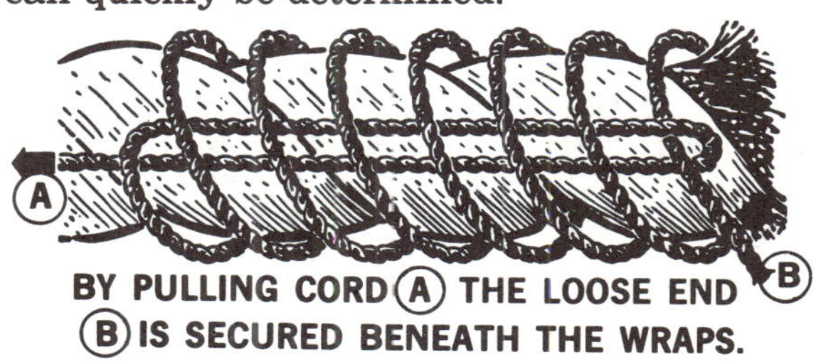

Figure 2—13 Technique used to whip end of rope.

Coiling. Two methods of coiling the rope are used—

One end of the rope is taken in the left hand; the right hand is run along the rope until both arms are outstretched. The hands are then brought together forming a loop which is laid in the left hand. This is repeated forming uniform loops until the rope is completely coiled. If there is any tendency for the rope to twist or form figure eights, it may be given a slight twist with the right hand to overcome this. *The rope should always be coiled in a clockwise direction.*

The rope is coiled around the left foot and knee while in the kneeling position with the left foreleg vertical. In coiling on the leg, start from the inside, bring the rope over the knee to the outside around the foot to the inside, and continue in this manner until the rope is coiled using the same technique as in the hand coil method.

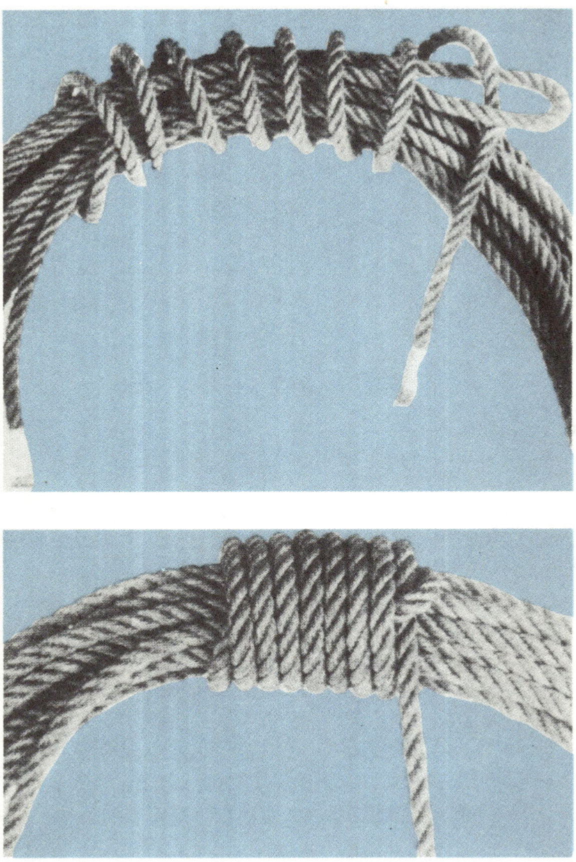

Figure 2—14 Tying of Coil.

Tying of the Coil. (fig. 2—14). In tying the coil, a 30 cm-long bight is made in the starting end of the rope and laid along the top of the coil. Uncoil the last loop and take the length of rope thus formed and wrap it around the coil and the bight. The first wrap is made at the open end of the bight in such a manner as to lock itself. Then continue wrapping toward the closed end until just enough rope remains to insert through the bight. Pull the running end of the bight to secure the wrapped rope. *A rope properly coiled has from six to eight wraps.* The coil can be carried either on the rucksack by forming a figure eight and doubling it, and placing the coil under the flap of the rucksack, or by placing it over one shoulder and under the opposite arm.

Rope Throwing. It will be necessary to carefully coil the rope before throwing. In throwing the full 36½ meter rope, grasp the coil in the right hand and take the end of the rope nearest the finger tips and anchor it. Take five or six loops from this end of the coil and hold in the left hand while retaining the remaining coil, which will be thrown first, in the right hand. A few preliminary swings will insure a smooth throw. The swings should be made with the arm nearly extended. *The coil should be thrown out and up.* A slight twist of the wrist so that the palm of the hand comes up as the rope is thrown will cause the coil to turn, the loops to spread, and the running end to fall free and away from the thrower. A smooth follow-through is essential. As soon as the coil is thrown and spreading, the few loops held in the left hand should be tossed out. Where possible, the rope should be thrown with the wind so that the running end is to the leeward side. As soon as the rope starts to leave the hand, the thrower shouts the warning "ROPE" to alert anyone below his position.

Knots. All knots used by the climber fall into four classes. They are—

- *Knots to tie the ends of two ropes together* (fig. 2—15).

  Square knot. The square knot is used to tie the ends of two ropes of equal diameter together and must be secured by a half hitch on each side of the knot.

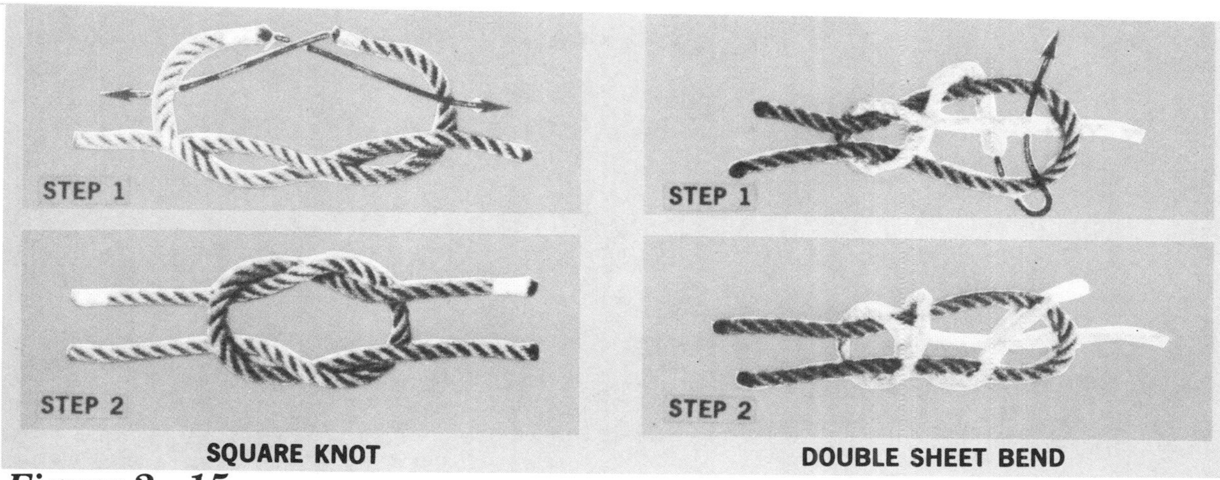

Figure 2—15

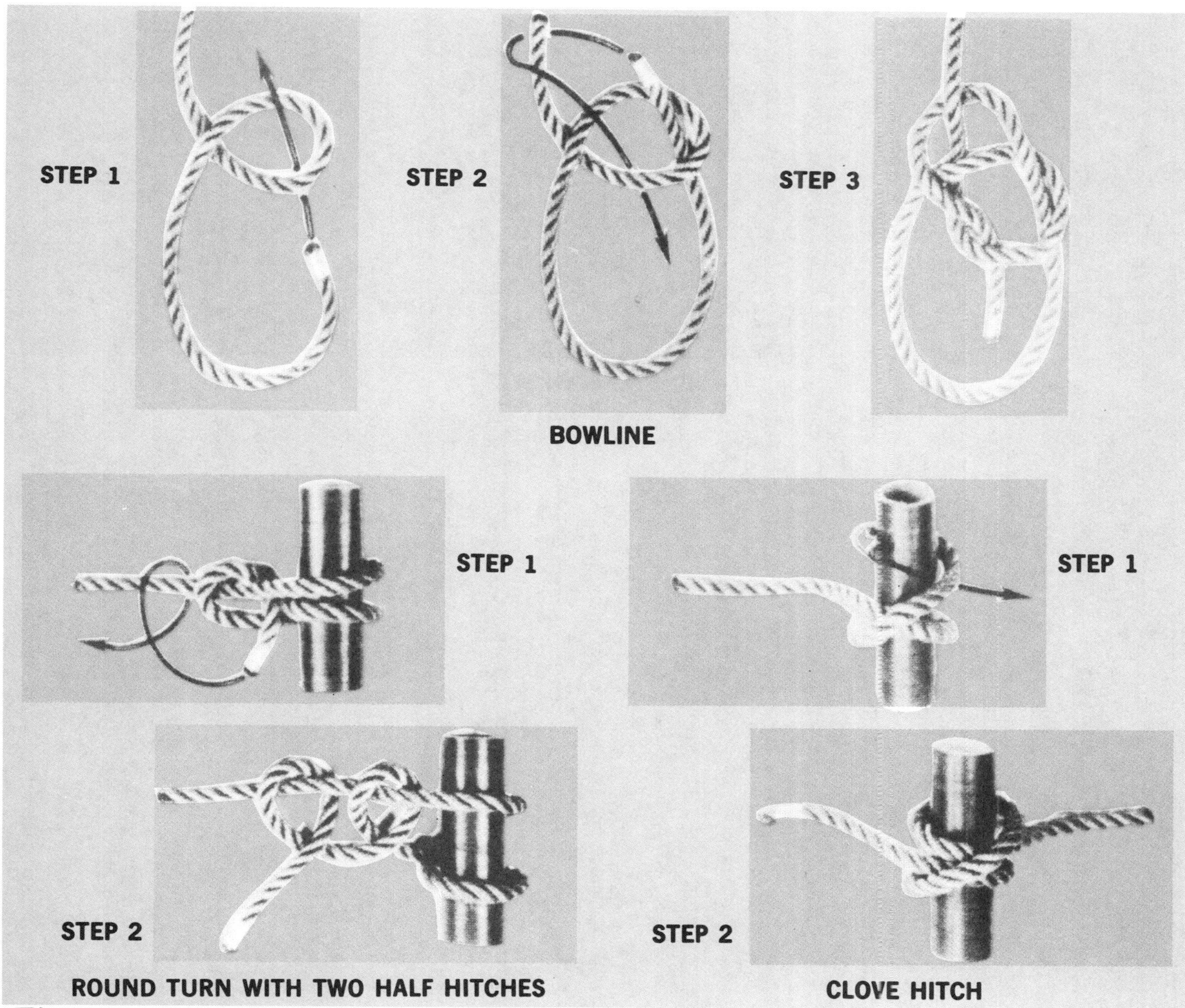

Figure 2—16

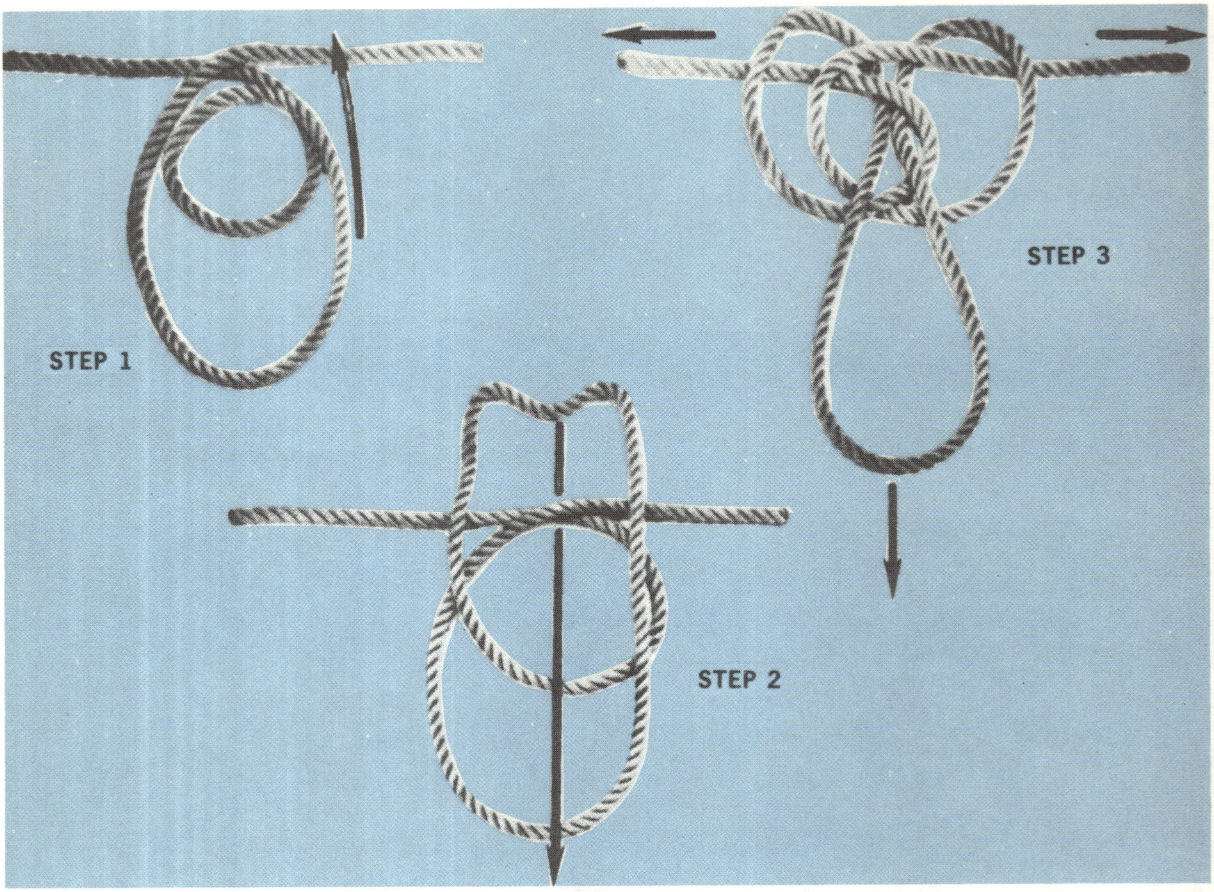

BUTTERFLY KNOT

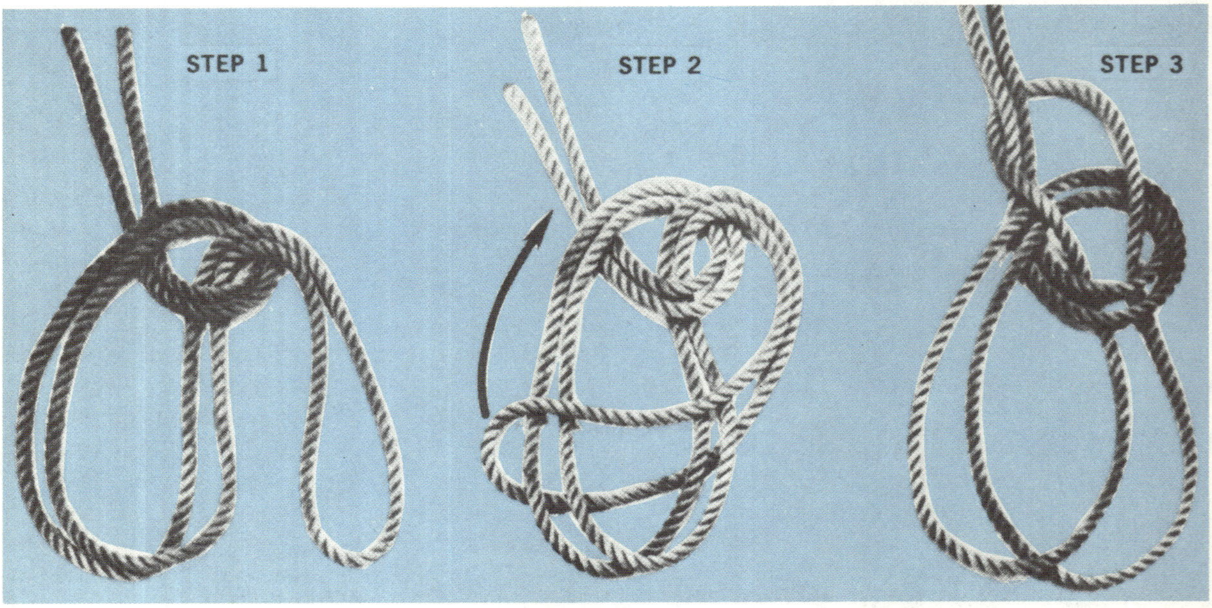

Figure 2—17 BOWLINE ON A BIGHT

Double sheet bend. The double sheet bend is used to tie together the ends of two ropes of equal or unequal diameter. It can also be used to tie the ends of several ropes to the end of one rope. When a single rope is tied to a multiple of ropes, the bight is formed with the multiple of ropes.

- *Anchor knots* (fig. 2—16). Anchor knots are used to tie the end of a rope to any object. An anchor knot is easy to tie and untie. Care must be asserted in selecting an anchor knot that will not work itself loose when alternate tension and slack are put on the rope. The round turn with two half hitches and the clove hitch may work loose under these conditions; the bowline will not. The three anchor knots are—

The Bowline. Used to tie the end man into a climbing rope and to tie a fixed loop in the end of a rope.

The round turn. The round turn is tied with two half hitches.

The clove hitch. Tension must be maintained to prevent slipping.

- *Middle rope knots* (fig. 2—17). Middle rope knots form a fixed loop or loops in the middle of a rope without using the ends. The butterfly knot is used for the middle man in a rope party and for tightening installation ropes.

The butterfly will form a single loop.

The bowline on a bight forms a double loop.

- *Special knots* (fig. 2—18). Examples of special type knots are—

Prusik knot. The prusik knot is tied with a short rope around a longer rope; for example, a sling rope around a climbing rope, in such a manner that the short rope will slide on the climbing rope if there is no tension applied, and will hold if tension is applied on the short rope. It is tied with a bight of rope or end of rope. When tied with an end of rope the knot is finished off with a bowline. The prusik knot is used to anchor a fixed rope to various anchors and in crevasse rescue methods.

Overhand knot. The overhand knot is used to make a knotted rope for a handline, to make the carrying rope for a suspension traverse, and stirrups in tension climbing. It is also used to temporarily whip the end of a rope.

Bowline on a coil. The bowline on a coil can be used by the first and last man on a climbing rope to take up extra and unnecessary slack. A half hitch must be employed behind the knot.

Three loop bowline. The three loop bowline will provide three bights, two of which can be adjusted against the other one. It is used mainly for a 4-piton anchor and in evacuation procedures.

Figure-of-eight slip knot. The figure-of-eight slip knot is used as an anchor knot on fixed ropes.

Transport knot. The transport knot is used to secure the tightening arrangement on rope installations and consists of a slip knot and half hitch.

Single and double retrievable bowline. These knots are tied to facilitate the retrieval of rope installations.

2—8. Pitons, Snaplink, and Piton Hammer

Purpose. In conjunction with rope climbing and rope installations, pitons are driven into cracks in the rock to provide—

- A secure point on the cliff to which the rope may be attached by means of a snaplink. If the leading climber falls, he may be held, pulley-wise, by a man below him. It also holds each following man. Successive pitons are driven as the climber moves upward.
- Secure points along the course of a fixed rope so that it will give greater support to troops using the rope.
- Artificial anchor points for rope installations.

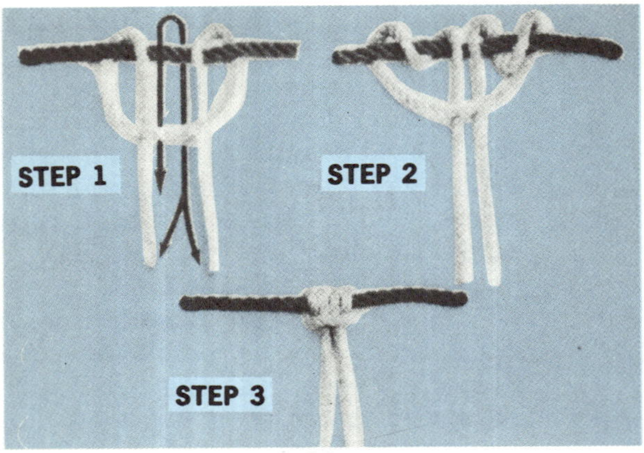

PRUSIK

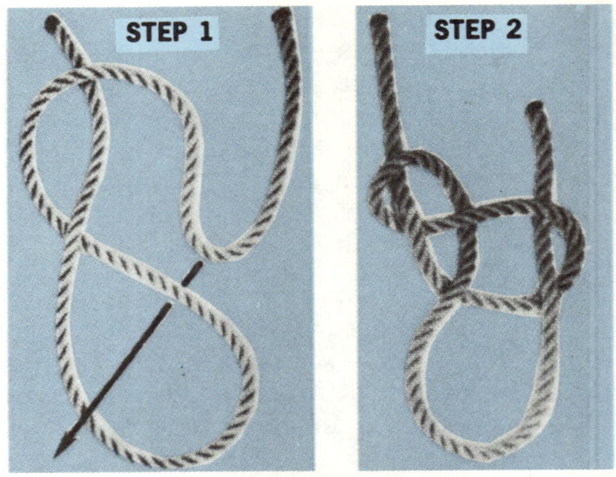

FIGURE-OF-EIGHT SLIP KNOT

OVERHAND

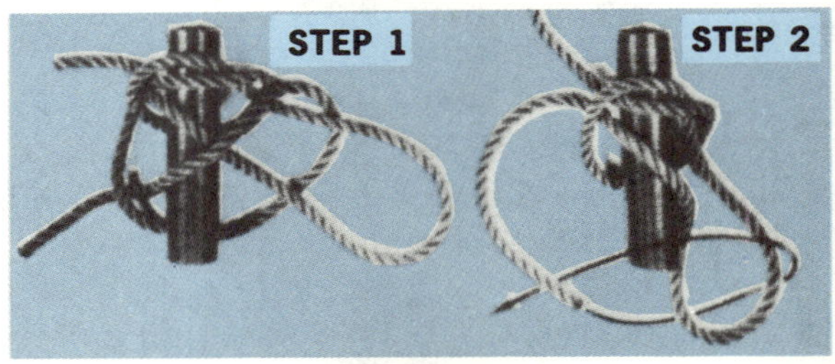

TRANSPORT

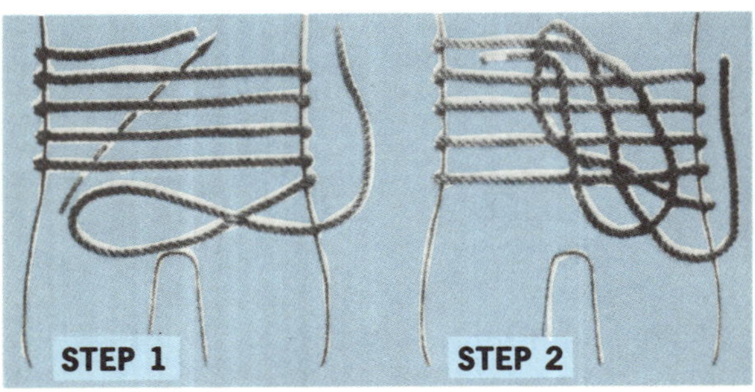

BOWLINE ON A COIL

SINGLE RETRIEVABLE BOWLINE

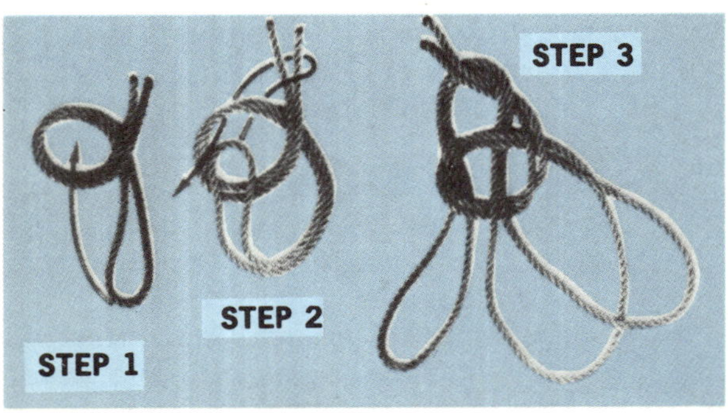

THREE-LOOP BOWLINE

DOUBLE RETRIEVABLE BOWLINE

Figure 2—18 ***Special knots.***

Pitons. (fig. 2—19). There are *four types* of pitons for rock:

- *Vertical,* for flush vertical cracks.
- *Horizontal,* for flush horizontal cracks and for offset horizontal or vertical cracks.
- *Angle,* for wide horizontal or vertical cracks. These must be placed with the wide or open side down in horizontal cracks and open side against either wall in vertical cracks.
- *Wafers* for shallow cracks, vertical or horizontal.

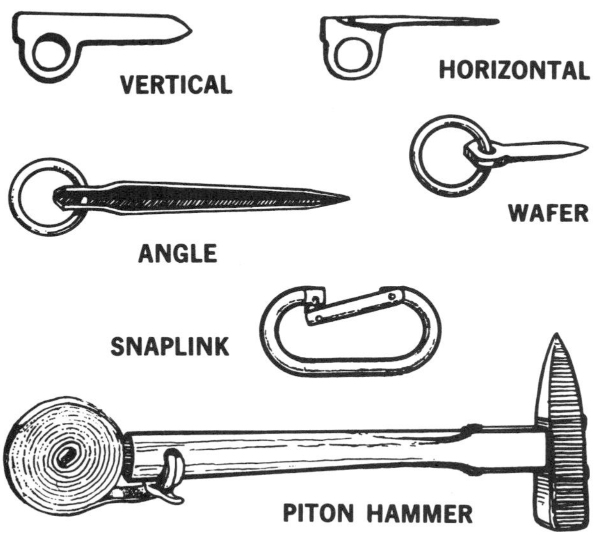

Figure 2—19 Pitons, piton hammer, and snaplink.

Placing of Pitons (fig. 2—20). Pitons are placed to increase the climbers safety. If well placed and tested they will limit his fall to twice the distance he is above the piton, plus the stretch factor of the rope. In placing pitons the climber should:

- *Study the rock.* See that driving of a piton will not split or weaken it. Test rock for soundness by tapping with the hammer. In hard, solid rock select a crack which is wide enough to admit 1/3 to 1/2 of the piton shaft before driving the piton in. The driving of pitons in soft or rotten rock is not always practical. When this type of rock must be used, loose rock from the crack should be removed before driving the piton. In this type of rock it is not necessary to be able to insert the piton into the crack as far as in solid rock. Select the right piton; the one that the rock will support best and that the snaplink can be hooked into after the piton is driven in.

- *Driving the pitons.* While driving watch the rock to see that it is not being weakened by further cracking. Watch the piton to see that it goes in smoothly and notice if the point hits a dead end. Listen to the piton's sound at each blow; good verticals and horizontals usually go in with a rising pitch; wafer and angle pitons will have no noticeable pitch so long as the ring is swinging free. Drive the piton with slow moderate strokes. *The piton should always be attached to the hammer thong or to the sling rope by a snaplink.* If the piton is then knocked out of the crack while it is being driven, it will not be lost. The greater the resistance overcome in driving the piton, the firmer it will be. An appropriately placed and well-driven piton into rock of average strength will withstand a force of from several hundred pounds (wafer) to more than 2,000 pounds (angle) exerted in the direction of pull.

- *Testing of pitons.* Pull up approximately one meter of slack on climbing rope, or use sling rope or piton hammer thong; snap rope into snaplink; grasp rope at least ½ meter from the snaplink. Making sure that you have secure footing, jerk vigorously outward, downward, and to each side, meanwhile observing the piton. Repeat if the test is questionable. Tap the piton. If the pitch has changed much, drive the piton in as far as possible; if the sound regains its original pitch the piton is good. If not, drive another piton in another place.

- *Snaplink.* The snaplink is used to fasten the rope to the piton, and will hold a load of 2,000 pounds when the gate is closed. The gate is the spring-loaded opening on the side of the snaplink and is used to facilitate attaching the snaplink to pitons, ropes, or other snaplinks. In snapping it into the piton, the climber should see that the snaplink will not cause unnecessary friction as he climbs beyond it, and that the gate is not likely to open accidentally because of pressure exerted by rock, rope, or piton.

- *Piton Hammer.* The piton hammer is used to drive pitons. The point is used for chipping rock or ice and cleaning out piton cracks, *and not for pulling out pitons.* The hammer is too lightly constructed to stand the stress. It is provided with a leather thong to secure the hammer to the body. When not in use the hammer is carried in the hip or breast pocket.

- *Second-Hand Pitons.* Pitons that have been used, removed, bent, and straightened *should be discarded.* In training areas, pitons already in place should not be trusted, inasmuch as weathering will loosen them in time. They should be tested and redriven until the climber is certain of their safety.

- *To Remove Pitons.* The climber should knock the piton back and forth in the crack with the piton hammer, and when they are loosened, pull them out with a bight of the climbing rope or sling rope, or hammer thong, which has been hooked into a snaplink. *It is advisable to be well braced when pulling out pitons,* as they sometimes tend to come out very suddenly. It is also advisable to see that no one is close enough to be struck by the extracted piton.

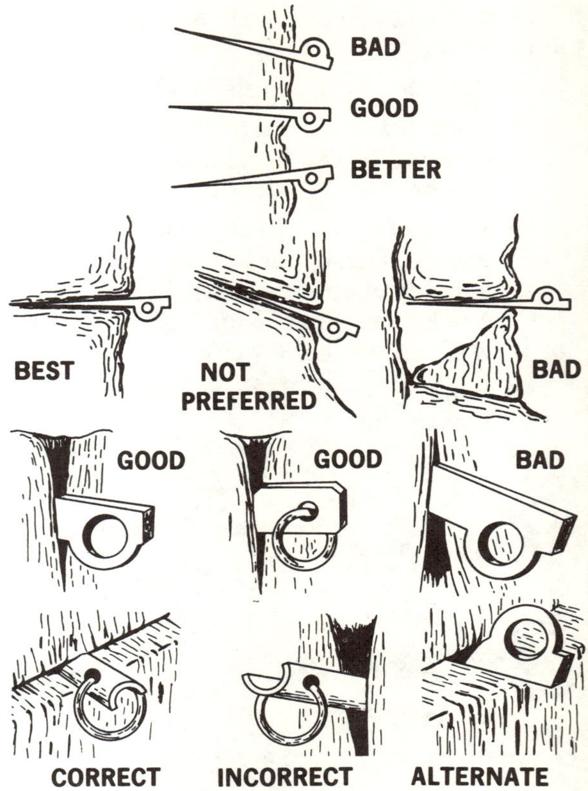

Figure 2—20 Placing of pitons.

2—9. Belays

Purpose. In party climbing, two or three men are tied in to the climbing rope. *Belaying provides the necessary safety factor or tension, enabling the party to climb with greater security.* Without belaying skill, the use of rope in party climbing is a hazard, not a help. When one man is climbing, he is belayed from above or below by another man who may use any one of several belaying positions. Belaying is also used to control descent on fixed installations. *A static belay* is one which allows no running of the rope. *A dynamic belay* is one in which the rope is allowed to run slightly and the fall is stopped more gradually.

Procedure for All Positions. The belayer must perform the following duties:

- Run the rope through his guiding hand, *which is the hand on the rope running to the climber,* and around his body to his braking hand, and make certain that it will slide readily.

- Anchor himself to a reliable rock projection or well placed piton with a portion of the climbing rope or his sling rope if his position is unsteady.
- Make sure remainder of the rope is laid out so as to run freely through the braking hand.
- See that the rope does not run over sharp edges of rock.
- Be *constantly alert* to the climber's movements in order to anticipate his needs. Avoid letting too much slack develop in the rope through constant use of the guiding hand. Keep all slack out of the rope leading to the climber, thus sensing his movement. Avoid taking up slack too suddenly to prevent throwing the climber off balance. When taking up slack the braking hand is not brought in front of the guiding hand, but just behind the guiding hand. This allows the braking hand to slide back and to remain constantly on the rope.
- *Brace well for the expected direction of pull in a fall,* so that the force of the pull will, whenever possible, pull the belay man more firmly into position. A climber should neither trust nor assume a belay position which he has not tested.
- Where necessary, seek a belay position that offers cover and concealment.

Be able, in case of a fall, to perform the following movements automatically—

- Relax the guiding hand.
- Apply immediate braking action. This is done by bringing the braking hand across the chest or in front of the body.

Belay Positions. There are three main types of belay positions—

- sitting belays
- standing hip belays
- piton belays

All of these may be made more secure when the belayer is tied with a section of the climbing rope or the sling rope into a belay. When the climbing rope is used, a clove hitch or a butterfly knot is tied into the rope which is secured with a snaplink and a piton. In using the sling rope, it is passed through the snaplink and around the belayer and then secured by a square knot. In both cases it is important that the anchor rope has the proper length without slack so that the belayer will not be pulled out of his position when holding a fall. The placement of the belay rope is determined by the directional pull of a possible fall of the belayed climber. If the pull is downward the rope is placed above the anchor rope and if the pull is upward the belay rope should be placed below the anchor rope.

- *Sitting belay.* (fig. 2—21). This is normally the most secure and preferred position. The belayer sits and attempts to get good triangular bracing by his legs and buttocks. Legs should be straight whenever possible, and the guiding hand must be on the side of the better braced leg. The rope should run around the hips. If the belay spot is back from the edge of a cliff, friction of the rope will be greater and will simplify the holding of a fall, but the direction of pull on the belayer will be directly outward. Make sure the rope does not pass over sharp edges.

Figure 2—21 Sitting belay.

398-868 O - 83 - 3

- *Standing hip belay*. (fig. 2—22). This is a weaker belay position and is only used where the sitting belay is not possible. An anchor for the belayer is essential. The belayer takes up a secure and well braced position. If possible, the back or shoulder is braced against the rock. The rope is placed around the back above the heavy portion of the hips and in front to the braking hand in such a way that the belayer is pulled more firmly into position when a fall is held. The climbing rope is placed either above or below the anchor rope according to the direction of pull of the belay rope.

Figure 2—22 Standing hip belay.

- *Piton belay*. (fig. 2—23). As soon as the leading climber has placed a reliable piton and placed his rope into a snaplink, which has been placed into the piton, the climber gives the signal "BELAY FOR PITON" to the belayer. The belayer should run the belay rope below his buttocks and a low anchor in line with the direction of pull. The braking arm should be extended. When a fall occurs, both knees should be bent to prevent the rope from sliding above the buttocks. As much rope as necessary is allowed to slide through the hand while the braking hand is brought in with steadily increasing resistance across the body in a dynamic belay. A fall is easier to hold with a piton belay than with a sitting or standing belay because of the added friction between rope, rock, and snaplinks. For this reason it is essential that the belayer use a dynamic belay to prevent the fall from being stopped too suddenly. When belaying a climber in a traverse which is near or horizontal to the belayer, the belayer must anticipate a pull sideways and straight forward, and should place the climbing rope above his buttocks and be anchored in.

Figure 2—23 Piton belay.

- *Use of natural anchors*. Where possible, the leading climber should pass his rope behind rock projections or trees which can give the same protection as a piton. He should avoid passing the rope over sharp edges or small cracks where the rope could jam or cause too much friction. When a rock or tree belay has been established, the climber should signal "BELAY FOR ROCK, OR TREE" as the case may be, and the belayer assumes a piton belay position. Likewise the belayer can increase his holding capability by placing the belay rope around a tree or rock projection in belaying directly to these natural objects.

Belaying. After the climber has taken up a belay position, he calls, "ON BELAY-TEST", which is answered by the man being belayed in calling "TESTING" at which time he puts his weight gradually on the rope in three progressively harder pulls without going off belay. Care must be taken that the rope is not jerked suddenly when starting the test, because this might pull the belayer out of position. The test should be made if possible in the direction of a possible fall. If, during the test, the position is found insecure, the belayer will signal "OFF BELAY". The climber must release all tension at once and maintain his belay position. The belayer must then find another position and have the test procedure repeated. When the position is satisfactory, the belayer calls "CLIMB". The climber answers him with the signal, "UP ROPE", in order to have the belayer take up the slack before he climbs. This procedure must be followed to prevent accidents. In training areas *all* commands should be prefaced with the climbers name for safety reasons. For example, a belayer may not be able to see his lead climber. He hears the command "ON BELAY-TEST" given by the leader of an adjacent party. Believing it to be his own leader, the belayer begins the testing procedure, promptly pulling his own leader, who is still climbing, off his holds.

2—10. Rappels

Purpose. The climber with a rope can descend quickly by means of a rappel—sliding down a rope which has been doubled around such rappel points as a tree, projecting rock, or several firm pitons secured one to the other with a sling rope.

Establishing a Rappel.

- In selecting the route, the climber should be sure the rope reaches the bottom or place from which further rappels or climbing can be done.
- The rappel point should be tested carefully, and inspected to see that the rope will run around it when one end is pulled from below and that the area is clear of loose rocks which may be pulled off.
- If a sling rope must be used for a rappel point, it should be tied twice to form two separate loops.
- The first man down should—

  Choose a smooth route for the rope which is free of sharp rocks.

  Place loose rocks, which the rope might later dislodge, far enough back on ledges to be out of the way.

  See that the rope will run freely around the rappel point when pulled from below.
- Each man down will give the signal "OFF RAPPEL", straighten the ropes, and see that the rope will run freely around its anchor. When silence is needed a prearranged signal of pulling on the rope is substituted for the vocal signal.
- When the last man is down, the rope is recovered. The climber should pull it smoothly, to prevent the rising end from whipping around the rope, and he should stand clear of falling rope and the rocks which may be dislodged by it.
- Inspect the rope frequently if a large number of men are rappelling on it.
- *Climbers must wear gloves for all type rappels to protect the palms from severe rope burns.*

The Body Rappel. (fig. 2—24). The climber faces the anchor point and straddles the rope. He then pulls the rope from behind, runs it around either hip, then diagonally across the chest and back over the opposite shoulder. From there the rope runs to the braking hand which is on the same side of the hip that the rope crosses; for example, the right hip to the left shoulder to the right hand. The climber should lead with the braking hand down and should face slightly sideways. The foot corresponding to the braking hand should precede the other at all times. He should keep the other hand on the rope above him just to guide himself and not to brake himself. *He must lean out at a sharp angle to the rock.* He should keep his legs well spread and relatively straight for lateral stability and his back straight since this reduces unnecessary friction. The collar should be turned up to prevent rope burns on the neck. Gloves should be worn and any other article of clothing may be used as padding for the shoulders and buttocks. To brake, lean back and face directly into the rock so that the feet are horizontal to the ground.

Figure 2—24 Body Rappel.

The Hasty Rappel. (fig. 2—25). Facing slightly sidewise to the anchor the climber places the ropes across his back. The hand nearest to the anchor is his guiding hand and the lower hand does the braking. To stop, the climber brings his braking hand across in front of his body locking the rope and at the same time turns to face up toward the anchor point. This rappel should be used only on moderate pitches. Its main advantage is that it is easier and faster than the other methods, especially when the rope is wet.

The Seat Rappels. The seat rappels differ from the body and hasty rappels in that the main friction is absorbed by a snaplink which is inserted in a sling rope seat fastened to the rappeller. This method provides a faster and less frictional descent than the other methods. *Gloves should be worn to prevent rope burns.* When using seat rappels at night it is advisable to use two snaplinks. Instead of running the rappel rope through the snaplink that is attached to the rappel seat, a second snaplink is inserted. The rope is run through the second snaplink in the usual manner. Opening the gate of the first snaplink and removing the second removes all tension from the rope and it is easily and rapidly disengaged from the snaplink.

Figure 2—25 Hasty rappel.

Attaching the seat. (fig. 2—26). The sling rope is placed across the back so that the midpoint (center) of its length is on the hip opposite to the hand that will be used for braking. An overhand knot is tied in front of the body. The ends of the rope are brought between the legs (front to rear), around the legs and over the hips and tied with a square knot and two half hitches on the side opposite the braking hand. The snaplink is placed through the single rope around the waist and through the two ropes forming the overhand knots. The snaplink is inserted with the gate down and the opening toward the body. The snaplink is then rotated one half turn so that the gate is up and opens away from the body.

STEP 1

STEP 2

STEP 3

Figure 2—26 Seat and placing of snaplink in seat.

- *Seat-shoulder method.* (fig. 2—27). In facing the rappel point, the rappeller snaps into the rope which passes up through the snaplink. The rope is then brought over one shoulder and back to the opposite hand (left shoulder to right hand). The same techniques are used in the descent as in the body rappel. This method is faster than the body rappel, less frictional and more efficient for men with packs and during night operations.

Figure 2—27 Seat-shoulder rappel.

- *Seat-hip method.* (fig. 2—28). The rappeller stands to one side of the rope (if right-handed stands to the left of the rappel rope facing the anchor, if left-handed to the right of the rappel rope). The rappel rope(s) are placed into the snaplink, some slack is taken between the snaplink and the anchor point, brought around the shaft of the snaplink and placed into the gate so that a round turn is made around the shaft of the snaplink. If a single rope is used this process is repeated to place two round turns around the shaft. Facing sideways, the climber descends using his upper hand as the guiding and the lower as the braking hand. The rope is grasped by the braking hand with the thumb pointing down and towards the body. The braking hand is held behind and slightly above the hip.

Braking action is obtained by closing the hand and pressing the rope against the body. *The rappeller should lean well out, at an angle from the rock, and make a smooth and even descent.* This method has the least friction and is fastest. However, special care must be taken that the rope is snapped correctly into the snaplink to avoid the possibility of the gate being opened by the rope. Loose clothing or equipment around the waist is apt to be pulled into the snaplink locking the rappel. For these reasons the rappeller must check carefully before each descent.

Figure 2—28. Seat-hip rappel.

2—11. Route Selection

Terrain must be analyzed if an efficient route is to be found through it. Every mountain reconnaissance unit should include climbers experienced enough to be able to estimate friendly or enemy capabilities in the region. Consideration should be given to the time element and skill of the troops involved, equipment available, and the support required. Probable effect of weather on the mission must be studied and care used in route selection to avoid danger of rock slides and other mountain hazards.

Scope. The military mountaineer must make detailed reconnaissance, noting with respect to each rock obstacle, the best approach, height, angle, type of rock, difficulty, distance between belay positions, concealment along the route, amount of equipment, and number of trained climbers needed to accomplish the mission on or beyond the rocks. If the strata dips toward the climber, holds will be difficult as the slope will be the wrong way. However, strata sloping away from the climber and toward the mountain mass will provide natural stairs, with good holds and ledges.

Vantage Points. At least two vantage points should be used, so that a three-dimensional understanding of the climb can be attained. Use of early morning or late afternoon light, with its longer shadows, is helpful in this respect. Actual ground reconnaissance should be made if possible.

Dangers to Avoid.

- On long routes, changing weather will be an important consideration. Wet or icy rock can make an otherwise easy route almost impassable; cold may reduce climbing efficiency, snow may cover holds; a cloudy night may be too dark for travel; a lifting fog may remove necessary concealment too soon. A weather forecast should be obtained if possible.
- Smooth rock slabs are treacherous, especially when wet or iced after a freezing rain. Ledges should then be sought.
- Rocks overgrown with moss, lichens, or grass become treacherous when wet. Cleated boots will then be far better than composition soles.

- Tufts of grass and small bushes that appear firm may be growing from loosely packed and unanchored soil, all of which may give way if the grass or bush is pulled upon. Grass and bushes should be used only for balance by touch or as push holds, never as pull holds.
- Gently inclined but smooth slopes of rock may be covered with pebbles that will roll treacherously underfoot.
- Ridges may be free of loose rock, but may be topped with the most unstable blocks in the region. Moreover, the climber is likely to skyline himself on a ridge. A route along the side of a ridge just below the top is usually best.
- Gullies provide the best defilade and often the easiest routes, but are more subject to ambush and to rock-falls from natural or enemy sources. The side of the gully is relatively free from this danger.
- Climbing slopes of talus, moraine, or other loose rock is not only tiring to the individual but dangerous, because of the hazard of rolling rocks to others in the party. Climbers should close up intervals when climbing simultaneously. Large groups should attempt to find numerous parallel routes.
- In electrical storms lightning can endanger the climber. Peaks, ridges, pinnacles, and lone trees should be avoided.

Rock Falls.

- Rock falls are the most common mountaineering danger. The most frequent causes of rock falls are—other climbers; great changes of temperature in high mountains and resultant splitting action by intermittent freezing and thawing; heavy rain; grazing animals; and enemy fire.
- Warning of a rock fall may be the cry "ROCK", a whistling sound, a grating, a thunderous crashing, or sparks where the rocks strike at night.
- A rock fall can be a single rock or a large rock slide covering a relatively large area.
- Rock falls occur on all steep slopes, particularly in gullies and chutes. Areas of frequent rock falls may be indicated by abundant fresh scars on the rock walls; fine dust on the talus piles; or lines, grooves, and rock-strewn areas on snow beneath cliffs.
- Immediate action is to seek cover, if possible. If there is not enough time to avoid the rock fall, the climber should lean into the slope to minimize his exposure.
- Rock fall danger is minimized by careful climbing and by judgment in choice of route.

Section III MOVEMENT OF TRAINED PERSONNEL

2—12. Unroped(Balance) Climbing

Much of the movement of a trained climber whose mission may be that of guide, observer, assault-team member, or security element will be without the aid or encumbrance of a rope. The terrain may not be difficult enough to require use of rope. Enemy action may require greater speed than will be possible with rope, the danger from enemy fire overbalancing the danger of unroped falls. Once the route for the main body has been selected and found relatively easy to travel, security elements may need to make only short scrambles to either flank. Balance climbing will sacrifice the safety that results from use of the rope. In terrain so rugged that precise routes must be followed, or on which fixed ropes would be desirable for inexperienced men but are not available, the trained climbers can be posted as guides at critical points to coach or otherwise assist untrained men over steep or hazardous portions.

2—13. Roped(Party) Climbing

Purpose. Trained climbers tie together two or three men to the climbing rope when, by so doing, they increase their mutual safety and ability on difficult rock. When one man is climbing he is belayed from above or below by another man. A two-man party is considerably faster than a three-man party. The strongest team is made up of two parties of two.

Procedure: Two-man party. One man chosen as leader because of his ability and experience will normally climb first. Both men tie in with a bowline on a coil. The second man takes a belay position and starts the test procedure. Having found his position satisfactory, he gives the order, "CLIMB." The leader will then climb to a suitable belay position. He should not take long leads, particularly where the climbing is difficult. If, because of the lack of a suitable belay position he has to take a long lead, or if climbing is precarious, the leader should use pitons for safety. In this case the belay man adapts his position to an upward pull of the rope. The use of too many pitons will cause excessive friction on the rope. The belayer should watch the slack and inform the climber by the call "6 METERS" when he estimates that there is only 6 meters of rope left. When the climber has reached a belay position, he calls "ON BELAY" and the test procedure follows.

While the second man climbs to reach the new belay position he removes any snaplinks. When the second man has reached the belay position he will change position with the belayer. The first man, or if more convenient, the climbers may bypass each other and lead alternately. This will speed up the climbing. The above procedure is repeated until the objective is reached.

Three-man party. In a three-man party each man has a number, the leader being number 1, the middle man is number 2, and the end man is number 3. The middle man must be an experienced belayer and the third man will usually be the least experienced man. If, as rarely occurs, a treacherous horizontal traverse must be negotiated, the least experienced man may be tied in the middle where he can be belayed from the man above him as well as from below. The signals are the same as for a two-man party, except that the number of the man involved must be used with the signal.

For example, the middle man may give the order, "Number 1 CLIMB" or "Number 3 CLIMB", to which the proper reply is "Number 1 CLIMBING" or "Number 3 CLIMBING". The leader climbs from the starting point to the first belay position, brings up number 2 and climbs up to a second belay position. He provides what security he can while number 2 brings up number 3. Number 2 then follows number 1, who then climbs to the third belay position. When not climbing, number 3 provides security either in the military sense or where desired as anchor man for number 2 when he is belaying. Similarly, number 1 man provides security while number 3 climbs. It must be remembered that no man climbs until so ordered by his belayer and *only one man climbs at a time.*

General rope signals. In addition to the belay signals, certain others are useful. The call "UP ROPE" is used when a climber discovers excess slack in his belay rope. The call "SLACK" is used when the rope is too tight for maneuver. If the climber is in trouble and wants a tight rope he calls "TENSION".

The call "FALLING" should be used to warn the belay man if the climber believes himself about to fall. When silence is necessary or a high wind distorts vocal signals, information can be conveyed by a prearranged system of jerks on the rope.

Descent. Where rappelling is not used, the party will descend in an order decided upon by the leader. If route selection is the primary consideration, the leader may want to go down first. If it is more important to provide belay security, he will go down last. More often than not the latter will be the case. Security for the last man down can be provided by the use of pitons.

Simultaneous movement. When the slope becomes gentler for a stretch, the movement of the entire party can be expedited by having all the men move at once. All men carry slack rope in neat coils which can be paid out or added to as the distance between the men

Figure 2—29 ***Direct aid climbing***

varies. The rope must be kept off of the ground but not taut.

Direct aid climbing. (fig. 2—29). In direct aid climbing the belayer holds the leader to the rocks by means of the climbing rope and pitons, the snaplinks in each piton serving as a pulley. Thus the leader is able to move up, sideways, and even up an overhang by driving successive pitons for tension, even though the natural holds are inadequate. Since direct aid climbing requires long experience and considerable equipment and is very exhausting to the climber, it should be used only as a last resort.

The direct aid climber should :

- Tie in with a bowline on a coil, finishing with a half hitch around the last single loop, leaving sufficient rope end to tie a bowline or figure-of-eight slip knot in the end of the climbing rope for a rest loop. A snaplink is placed in this loop and fastened to the piton in front of the climber to give him security while he advances the rope. He should rely on the rest loop to preclude exhausting the belayer on a long direct aid climb.
- Drive pitons no closer together than is necessary for safety.
- Never hook a finger in a piton, but put in snaplinks and climbing ropes and hold them.
- Use stirrups made from sling ropes for foot holds if their use is deemed necessary. To form a stirrup, the ends of the sling rope are tied together with a square knot. An overhand knot is tied in the middle of the loop thus formed but a little off center so that the knot at the bottom is to one side. Another overhand knot is tied at the top or opposite end from the joining knot and is again offset so that this loop hangs open. A snaplink is then inserted through the small top loop.
- Rely on the belayer's tension, keeping both hands free for the work above.
- Use alternate (two-rope) tension for the most difficult pitches. This is easier if there are two belayers. However, if there is only one belayer, the following method may be used: The rope from the leader's waist passes around the belayer's waist. The second rope, fastened around the leader's hips runs to the same side of the belayer as the first rope, and beneath his hips. The leader climbs, the belayer holding him to the cliff with tension first on one rope and then on the other, always keeping one of the ropes slack so that the leader can snap into the next higher piton.
- The belayer should anticipate the leader's needs so that few signals need be given.
- When the leader needs tension to climb a pitch, the second man will also need assistance on the pitch. This is furnished by rope slings left in the pitons or, if alternate tension has been used, one of the two ropes can be tied at the top as a handline.

Section IV UNIT MOVEMENT

2—14. Anchors

The basic elements for the techniques applicable to movement are :

Anchors. In the setting up of all rope installations the problem of the main anchor is a great one. The ideal situation is to have some good natural object such as a firmly rooted tree or solid rock projection. Since this is not always available, anchors must be made or devised by artificial means. This type anchor is called a "dead-man" and the leader of the installing party must decide which form is most efficient in regards to speed of installation, safety, and durability.

Natural anchors. (fig. 2—30). Since these are always preferable, their use should be studied with care. If a tree is to be used, its firmness is of the greatest importance. This is especially true if the installation will be used for any length of time. Trees growing on rocky terrain should be treated with suspicion, since their roots normally are shallow and spread out along a relatively flat surface. If a tree has been found satisfactory, the rope may be tied to it with any of the anchor knots. If the installation is to be used for a great length of time, the bowline or the round turn with two half hitches is preferable to the clove hitch. When the bowline is used, a round turn should first be made around the anchor, (before the bowline is tied), to prevent it from "seizing up" and becoming very difficult to untie. If rock projections can be used, their firmness is again of primary importance. They should be checked for cracks or any other signs of weathering that may impair their firmness to any extent. If any of these signs exist, its use should be avoided except in cases of absolute necessity, and then only after careful and thorough testing. Sharp edges will almost always be found on rocks and these should be padded carefully with extra clothing, rags, branches, or grass.

Figure 2—30 ***Natural anchors***

Artificial anchors. (fig. 2—31). Artificial anchors can be divided into two main classes. These are anchors that are installed in earth or dirt and those that are put on rock with pitons.

- Artificial anchors in earth are of two types. The single timber deadman is the safest type, although its construction requires considerable effort. A trench 2 meters long and 1 meter deep and wide enough to work in should be slanted so that it is at right angles to the direction of the pull. The side of the trench towards the strain should be slanted so that it is at right angles to the pull. Another trench about 30 cm wide is dug so that it intersects the main trench at a 90° angle in the middle. The bottom of this trench would be parallel to the strain and should meet the bottom of the main trench. A log 2 to 3 meters long and 30 cm minimum in diameter is normally used for the deadman. The log is then put into the main trench and covered with dirt with the exception of the part adjoining the second trench. If the dirt is not firm, stakes the same length as the depth of the trench should be placed between the deadman and the slanted side in an upright position.
- The *picket holdfast* is easier to construct, but will not hold as much strain as the deadman. Pickets 2 meters long and 8 cm in diameter, are driven into the ground a depth of 1 meter, one

behind the other 90° to the direction at pull. The head of each picket, except the last one, is secured by lashing it to the base of the next one in line. The lashings should be tight as possible and racked. The pickets should be driven at right angles to the line of pull and the distance between the pickets should be one to two meters. The anchor line is tied at the base of the picket closest to the load.

- The *four piton anchor* is the least desirable of all anchors and its use should be avoided whenever possible. At least four pitons are driven firmly and solidly into the rock. In the rope, one less bight than pitons are tied in such a manner as to leave a running end of at least 5 meters (for instance, if four pitons are used, a three-loop bowline will provide the necessary bights). Snaplinks are placed in all the pitons and bights. Starting with a piton on either extreme end, the running end is placed in the snaplinks alternately between the pitons and bights. It is then tied in back to the standing part. This anchor must be checked frequently. If one piton should pull out, the strain will be equalized between the others. At the first opportunity the piton anchor should be set up again.

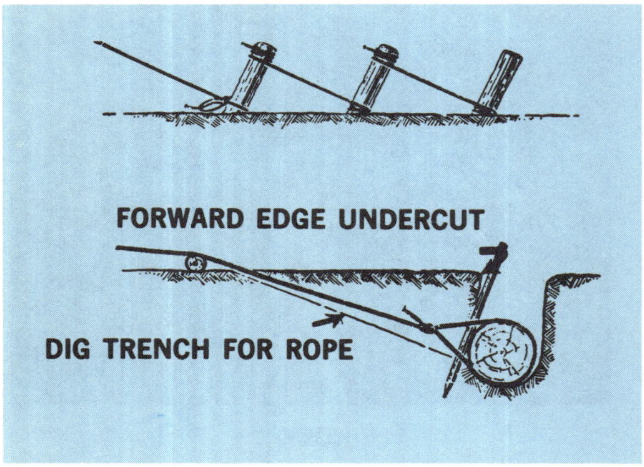

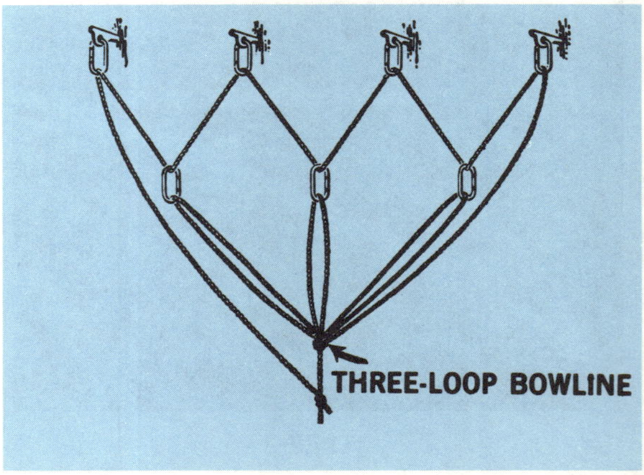

Figure 2—31 Artificial anchors

Tightening Knots. Are used for tightening fixed ropes, suspension traverses, or any other similar installation.

- *Transport tightening system.* A butterfly knot is tied in the static line far enough in front of the anchor to allow for tightening of the rope with the bight of the butterfly approximately 30 cm long. This knot should also be placed so that it acts as a safety factor for the man descending, preventing him from hitting the lower anchor. A pulley effect, for tightening the static line is obtained by inserting a snaplink into the butterfly, passing the running end around the anchor, and inserting it through the snaplink. The transport knot is made by tying a slip knot secured with a half hitch.
- *Prusik tightening system.* (fig. 2—32). A butterfly knot is tied in the static line far enough in front of the anchor to allow for tightening of the rope, with the bight of the butterfly approximately 30 cm long. This knot should also be placed so that it acts as a safety factor for the man descending, preventing him from hitting the lower anchor. A pulley effect, for tightening the static line, is obtained by inserting a snaplink into the upper loop of the butterfly, passing the running end around the anchor and through a prusik knot in the bight of

the butterfly, and finally through the snaplink. The prusik knot, as the static line is tightening through use of mechanical advantage, acts to cinch the tightening rope. The running end is then secured with a half hitch on the rope opposite to the prusik knot (fig. 2—33). The prusik tightening knot is normally used when there are not enough personnel to use the transport knot.

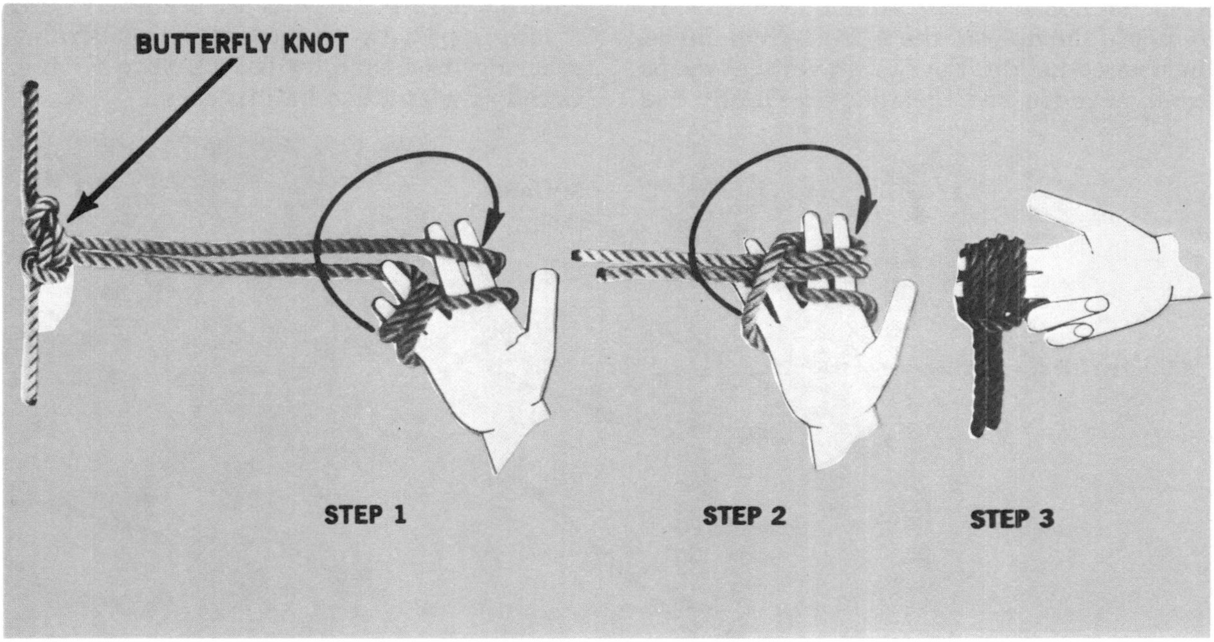

** Figure 2—32 Prusik tightening knot.*

**Figure 2—33 Use of butterfly knot and prusik knot in tightening.*

A-frame. (fig. 2—34). An A-frame is constructed in the following manner; after selecting two sturdy 3 to 4 meter long poles, two sling ropes are tied together and secured to the poles (approximately 1 meter from the top), with a clove hitch on one pole. Six to eight horizontal wraps are made downward to insure the apex of the A-frame remains at the desired height. Four to six vertical wraps are then made and the ends are finally tied tightly with a square knot and finished off with half hitches. Make sure the square knot is not tied in the crotch formed by the two poles. When the bottoms of the poles are spread apart, the resulting bipod forms the A-frame. To keep the bottom of the A-frame from spreading, the poles are braced by tying a sling rope between the legs of the A-frame, securing it on each leg with a clove hitch or round turn and two half hitches.

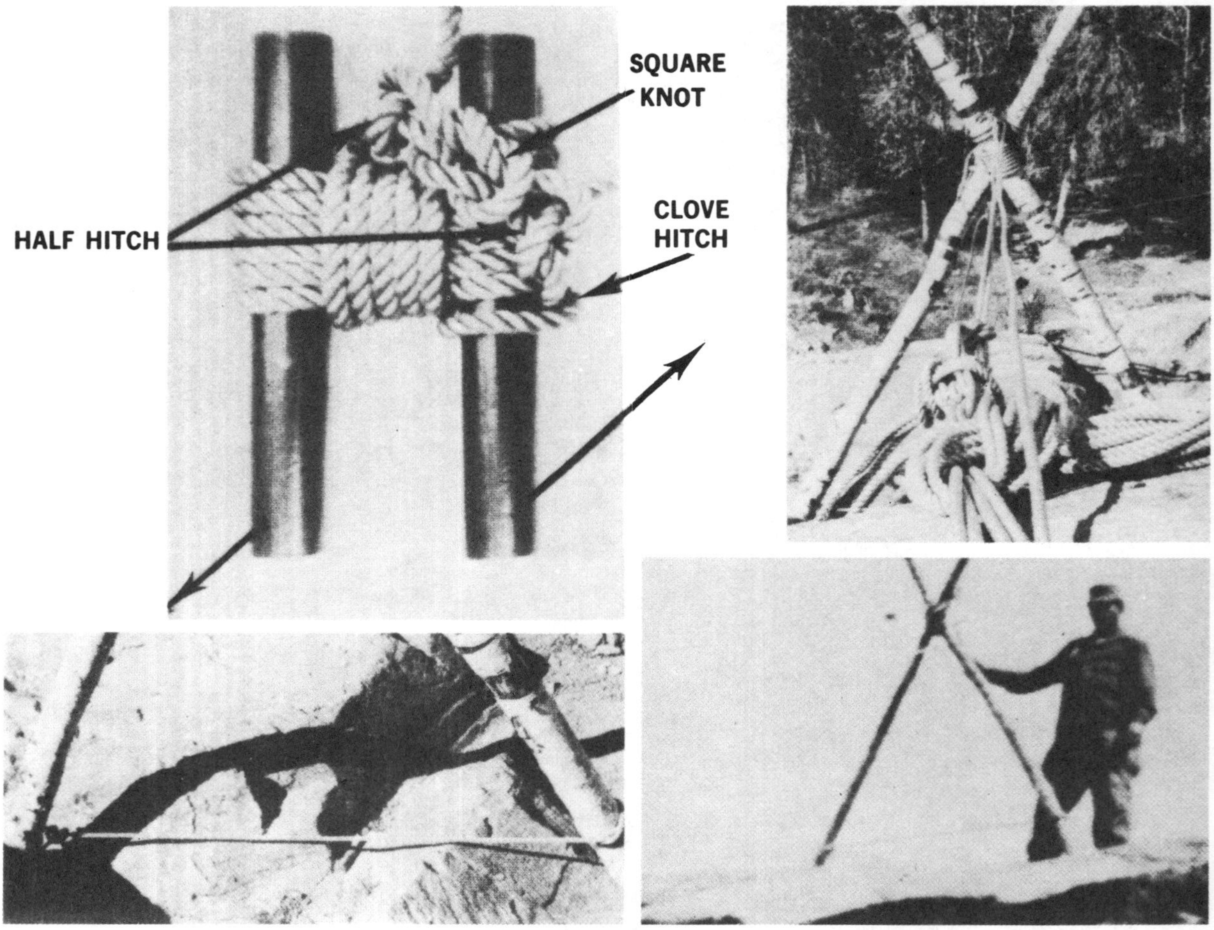

Figure 2—34 ***A-frame.***

2—15. Fixed Ropes

Once troops are familiar with the basic elements outlined in paragraph 2—12, the following fixed ropes may be installed to assist individuals and units in moving over difficult terrain with their equipment, weapons, and supplies.

Simple Fixed Rope. (fig. 2—35). A simple fixed rope is made by anchoring one end of a climbing rope and using the line thus formed to aid in climbing. The procedure is as follows: Number 1 ties in, plans route, and climbs on signal from number 2 who belays or sees that the rope runs free. At the top of the pitch number 1 takes up slack, either ties the rope to an anchor point, or gets into a body belay. After installation, the unskilled men then climb, using the rope for all desired aid. Only one man climbs at a time, and signals to the next man, "CLIMB", when he has reached the top. In difficult places overhand knots can be tied into the rope to assist the men climbing. The last man up retrieves the rope and coils it.

Figure 2—35 ***Simple fixed rope.***

Fixed Ropes. (fig. 2—36). Fixed ropes are installed by trained military mountaineering teams and are used to assist men with heavy loads over difficult terrain, safely and quickly. This type of installation differs from the simple fixed rope in that it employs many anchor points and is of a more permanent nature. To install—number 1 ties into the leading end of the rope and moves up the selected route threading the rope through snaplinks inserted in pitons he has driven, or sling ropes, tied to natural anchors. Number 2 remains at the starting point, belays number 1, and sees that the rope runs freely; warning the climber when only 6 meters of rope remains. Should more rope be required, another climbing rope is tied to the first rope. When the fixed rope is installed from the top down, number 1 man secures the rope with an anchor knot to the top anchor. If descending, he then ties the fixed rope to the intermediate anchor points, tightening the rope with prusik knots to anchored sling ropes, and the figure-of-eight slip knots into snaplinks while he is still belayed by the number 2 man. As he reaches the lower anchor point the fixed rope is then finished off with the transport tightening system.

When the fixed rope is secured from the bottom up, the number 2 man anchors the rope at the bottom anchor while the number 1 man secures the rope temporarily to the upper anchor. As the number 2 man moves up to the number 1 man, he anchors the rope to the intermediate anchor points and the rope is then secured with a transport tightening knot at the upper anchor. When the installation team consists of several members, they assist in securing the fixed rope to the intermediate anchor points while the number 2 man moves up to join the number 1 man. In this manner each section between anchors is tightened independently. The prusik knots and the slip knots in the snaplinks serve both to increase the tension of the rope and to make each section independent. This is a safety factor—a break in one section of the rope will not affect another section. The prusik knot or slip knot should be tied slightly above the anchor point. The transport tightening knot is used to finish off the fixed rope. Emphasis should be placed on the use of natural anchor points throughout the fixed rope and care taken in its installation to obviate the necessity of troops crossing it at any point. When snaplinks are unavailable, sling ropes in pitons may be used.

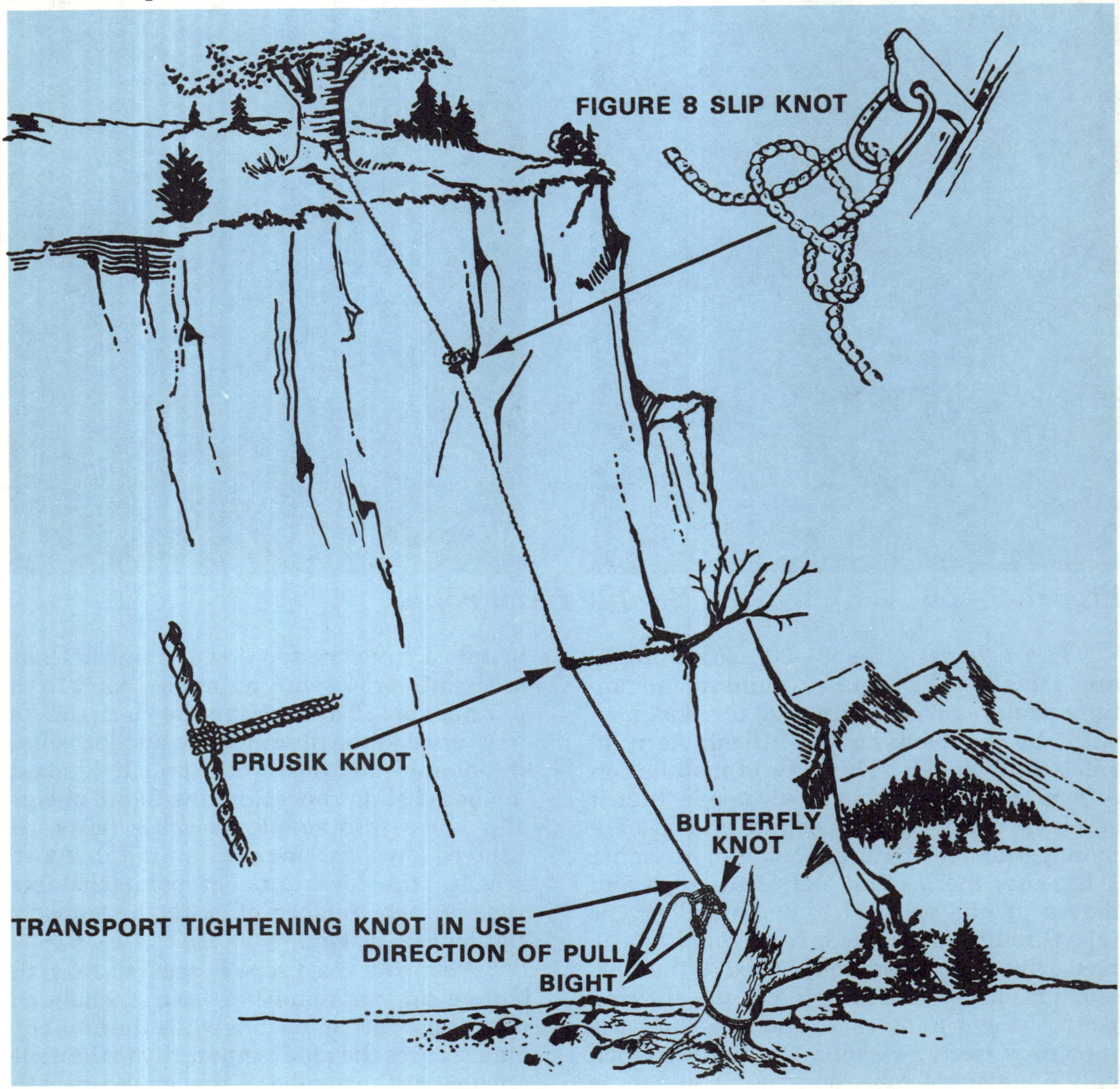

Figure 2—36** ***Fixed rope.

Maintenance of the fixed rope is the responsibility of the team that installed it. It is also their duty to police the route, removing loose rocks, brush, and any other hindrance to the rope and troops, and to protect the rope with padding where necessary. A diagonal or traversing route should be chosen for reasons of safety. The route should not be so difficult that the troops arrive at the top too exhausted to complete their missions. Troops moving over a fixed rope must remember and use all the techniques of mountain walking and climbing. Use the fixed rope for direct aid only when necessary, but always have at least one hand on the rope. On difficult pitches employ friction footholds by leaning back and climbing hand over hand on the rope.

2—16. Hauling Lines

When inexperienced men carrying heavy loads are required to climb fixed ropes, it may become necessary to set up a hauling line or a suspension traverse. Expedients may often serve well for all described equipment except the rope.

Vertical Hauling Lines. A vertical hauling line is an installation for moving men or equipment up vertical, or near vertical pitches (fig. 2—37). It is often used in conjunction with a fixed rope where the fixed rope is used for troop movement and as a hauling line for their equipment. Generally three climbing ropes, sufficient snaplinks, and the equipment for construction of an A-frame are necessary for this installation, but any expedients may be used that will aid the construction.

To install, select a route which has a good top anchor point, a natural loading and unloading platform at the bottom and the top, and which affords sufficient clearance for easy hauling of troops or equipment. The ideal platform at the top will allow construction of the vertical hauling line without use of an A-frame, otherwise the construction will be as follows: The anchor rope is doubled and the end with the bight placed between the top V of the A-frame so that a 30 cm long loop hangs down. A clove hitch is tied on each pole and the ropes tied to the same or different

Figure 2—37 Vertical hauling line.

anchor, passed through the top of the V of the A-frame, then knotted with overhand knots evenly spaced approximately 20 to 30 cm apart, and then lowered. This is the rope used by the troops as a simple fixed rope as they are being hauled up the pitch. Two snaplinks with gates opposed are inserted in the loop hanging from the A-frame to form a pulley. A pulley rope is formed by tying the ends of a climbing rope with a square knot and half hitches, and inserting it into the snaplink.

There are several ways of tying into the pulley rope—with a sling rope seat attached to the pulley rope with a snaplink inserted in a butterfly knot, or with two butterfly knots tied so that when they are passed around the body and connected with a snaplink, a belt is

398-868 O - 83 - 4

formed. If only two climbing ropes are available, the anchor rope and knotted rope may be combined, and a double sling rope used for the pulley loop. If equipment only is being hauled up, it is not necessary to use the knotted rope, but it may be necessary to use a belay rope. To move materials or troops up on one side of the hauling line, the other side is pulled from below.

Troops using the hauling line for movement must employ all applicable principles of climbing. Two operators should *always* be stationed at the top of the vertical hauling line to aid men or retrieve loads when they reach the top. They tie in with a bowline around their waist. The opposite end of this sling is secured to one of the anchor ropes with an end-of-the-rope prusik so that the operators are free to slide this knot back and forth for freedom of movement while operating the vertical hauling line. In use, the A-frame should lean slightly over the edge of the cliff to prevent excessive wear on the ropes passing over sharp rocks.

Suspension Traverse. The suspension traverse is used to move men and equipment, over rivers, ravines, chasms, and up or down cliffs (fig. 2—38). The traverse may be made on a plane varying from the horizontal to the near vertical. To install the traverse a suitable route must be reconnoitered with emphasis placed on the location of suitable anchor points. These must be of sufficient height to allow clearance of loads being transported over any obstacles. The static line must be carried to the upper or farther anchor by a single man, or a climbing party, and secured. It is then secured to the lower anchor and tightened by use of a tightening knot. If nylon climbing rope is to be used as a static line, it should be used double to minimize sagging and increase the safety margin. A 25mm manila rope should be used for installations which get heavy use. A carrying rope is made by tying the ends of a sling rope with a square knot, and half hitches, tying an overhand knot into the loop above the square knot and tying a second

Figure 2—38 ***Suspension traverse.***

overhand knot slightly below it. The belay rope is tied to the center of the three loops of the carrying rope and then a snaplink is inserted into the upper small loop and into the static line (fig. 2—39). To secure either the loads or the men to the static line this carrying rope is merely passed through the seat or lashings and the large loop snapped into the snaplink on the static line. When the man or load has been hooked onto the static line, a belayer lowers the load by using a body belay, belaying through a snaplink, or around a tree, and stops him gradually as he nears the bottom. If it is a steep traverse, and the descent is rapid, protective padding must be used by the belayer to reduce the friction on the body. If the load is being raised or the traverse is horizontal, it will be necessary to have a belay rope from both anchor points and to pull the load up or across and retrieve the carrying rope.

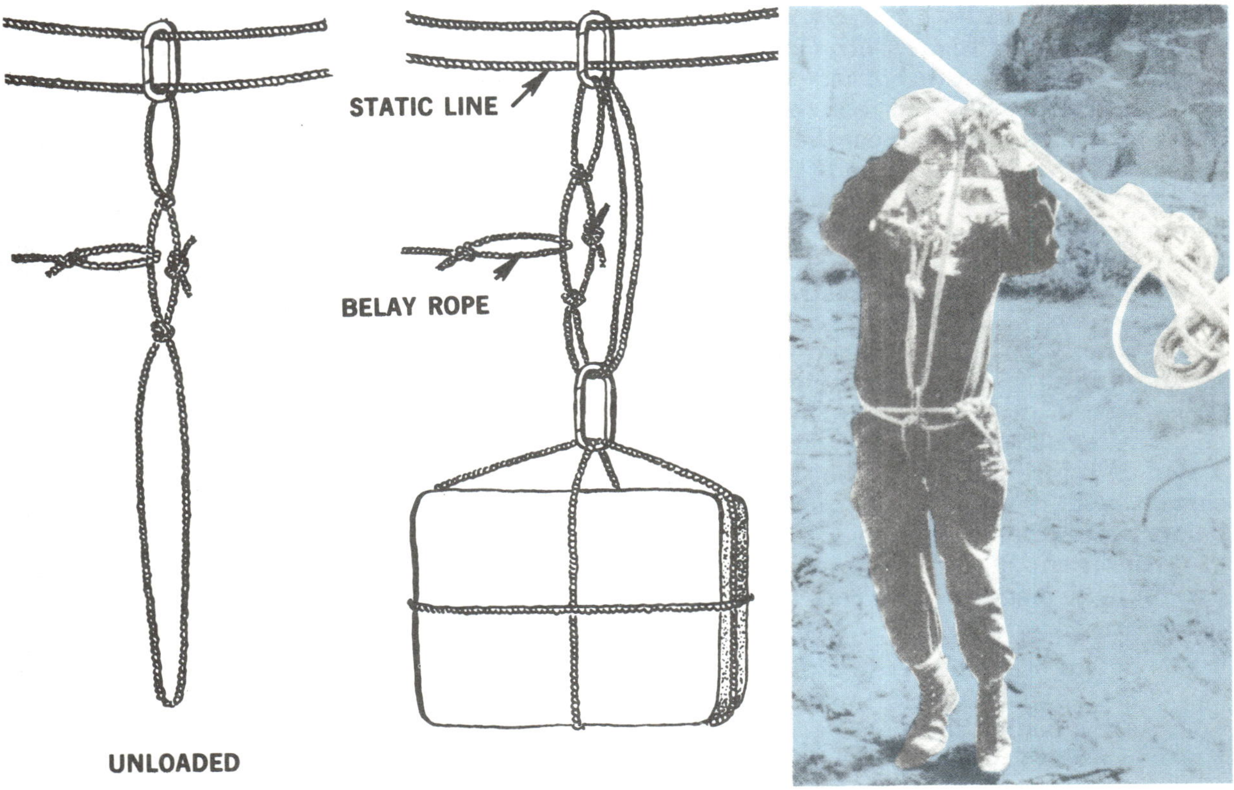

Figure 2—39 Carrying rope for use on traverse.

2—17. Mountain Stream Crossings

The techniques of the trained climber may be used in crossing mountain streams. Sudden rains or thaws can change placid streams into roaring torrents. These may have to be crossed without special equipment or the assistance of specialized troops. All mountain soldiers should be trained in crossing mountain streams.

The best time for crossing is in the early morning when the water is low. As glaciers, snow, or ice melt during the day, the rivers rise, reaching their maximum height between midafternoon and late evening, depending on the distance from the source.

Prior reconnaissance of possible crossings must be made before the arrival of the main body, so that the best place for the crossing may be selected. A crossing point should be chosen, if possible at the widest point of the stream where the water is normally not as swift and usually not as deep, or where it branches into several smaller streams. Cover should be available on the banks. Wherever possible, select a point where there are few, if any, large stones in the river bed. Water is normally turbulent over large stones, while it flows smoothly over small stones. Large stones increase the difficulty of maintaining an easy footing. The place for crossing should be clearly marked on both banks.

Shoes should be worn to prevent foot injuries, but socks and insoles should be removed and kept dry.

A shallow stream with a moderate current can be forded without the use of ropes or logs. The heavier and more experienced men are dispersed throughout the crossing to assist the others crossing, and to help break up the current. Experienced men, who are capable swimmers, should be posted downstream from the crossing site. The river should be crossed at an upstream angle, facing upstream. The feet should be wide apart, and kept flat with the bed of the stream, and should always be set down on the upper side of any obstruction in the stream bed. The legs should be dragged through the water, not lifted, so that the force of the current will not throw the individual off balance and drag him under.

Swifter and deeper rivers may be crossed by groups of men assisting each other.

- To cross using the ring method, a squad of men form in a ring, locking hands with each other and placing the arms behind their backs. The body is bent well forward.
- In the chain crossing, a squad of men form in line, lock arms with each other, and then cross their own arms and lock their hands in front to give added support. The line then moves across diagonally downstream.
- Use of bridge and stream crossing expedients are covered in paragraph 2—18.

Wherever possible and when the degree of experience permits, torrential streams should be forded individually in order to effect speedier crossing. If the current is moderate and the water less than knee deep, a staff or a walking stick may be used and worked ahead of the individual on the downstream side. Only experienced soldiers should attempt an individual crossing of a strong, deep current, and then they should proceed with utmost caution. Where the current is fast, the water is more than knee deep, or when it has a shifting bottom; crossing is facilitated by using a rope or balance pole that is at least head high. The balance pole is worked ahead of the individual, always on the upstream side. The weight of the body must be evenly distributed between the pole and the feet to maintain the necessary points of support. The pole is first moved forward and planted, and the feet are then moved ahead.

- Rapids or a fast-moving deep stream filled with large boulders may be crossed by jumping from stone to stone. Logs or trees may be felled from the banks to facilitate the movement. *As a safety measure a climbing rope should be used during the crossing.* Individuals tie themselves in with a bowline and are belayed by the security man who must pay the rope out gradually in order that the movements of the men will not be impeded. The hip belay may be used by the security man.
- When jumping from boulder to boulder, the soldier should jump from a crouching position pushing off simultaneously with both feet, and landing with both feet flat on the rock.
- The first man to cross takes up a good belay position and helps to insure the safety of those following.
- If the current of a river is rapid and deep, and yet can be crossed by using a rope support, a handline is strung from bank to bank. The most experienced individual ties into a rope and is belayed across the stream. The rope is then tied into whatever supports can be found on each bank and the unit crosses in column.
- When crossing, it is imperative that each individual stay on the

downstream side of the rope, as the current has a tendency to pull one under the rope.

- If equipment is being carried, it is advisable to use a sling rope around the individual's waist and attached to the handline with a snaplink.
- For carrying an injured man across, a rope seat may be made and attached to the individual who is to carry the injured man. The injured man is then hoisted to the back of the carrier and fitted into the rope seat. For additional security, a snaplink attachment is made to the hand line from a rope around the carrier's waist. The carrier then makes the crossing as prescribed for the other individuals.
- Rescue posts are set up at various points downriver on both banks. These posts should be manned with best men, equipped with ropes and poles. If the current sweeps a man off his feet and downstream, he is then rescued at the rescue posts.
- The last man to cross removes the rope from its belay points, and is belayed across the river from the opposite bank.

Where powerful torrents of water between sheer cliffs are encountered, a suspension traverse should be used.

- The individual selected to take the rope across must be the best trained, most experienced man in the unit. The man crossing uses a balance pole and a rope as previously described. The stronger the current, the greater is the angle at which the pole must be placed in the water.
- The crossing is assisted by the belayer who pays out the rope as the crossing man advances. If there are large boulders protruding from the river and lying in or near the path of the man crossing, he should utilize the rocks by crossing so that the rope falls against the upstream sides of the stones, thus affording himself an intermediate belay.
- As soon as the crossing man arrives on the opposite bank, he ties the rope to a solid anchor.
- Once the rope is tied in, it is tightened from bank to bank so that it is several meters above the water.
- Individual rope seats such as are used in rappelling with a snaplink may be made up for each man. The snaplink in front of the rope seat is then snapped into the crossing rope and the individual swings out, pulling himself across. Only one man should cross at a time.

2—18. Bridge and Stream Crossings

Special bridges and other crossing expedients useful in mountain operations are—

Branches Covered With Snow. Snowshoe, ski, and foot troops can use branches covered with snow as a crossing expedient to traverse shallow streams.

Ice. Ice bridges can be constructed and used as a crossing expedient. For details of construction see FM 31—71 and TM 5—349.

Rope Crossings. There are three accepted rope crossing methods, all of which require a climber to establish the anchor on the other side of an obstacle.

- *One-rope bridge.* (fig. 2—40). The one-rope bridge is constructed using the standard climbing rope. Because of the stretch factor in nylon rope, gaps exceeding 20 meters should not be bridged utilizing this method. The rope is anchored with a round turn and two half hitches, or bowline, and is tied off at the other end with a transport knot. The method used to cross the one-rope bridge is known as the horizontal traverse. This traverse can be accomplished by three means—

  The crosser lies on top of the rope with the left instep hooked on the rope and the left knee bent, the right leg hangs straight to maintain balance. Progress is made by pulling with the hands and arms.

  The crosser hangs below the rope suspended by the hands and heels crossed over the top of the rope.

In the final approach, the same method is employed except the crosser ties a rappel seat and ties into the rope with a snaplink. *Of the three methods employed, this is safer and preferred.*

- *Two-rope bridge.* (fig. 2—41). This bridge is constructed similarly to the one-rope bridge, except that two ropes are used. The ropes are spaced approximately 1.5 meters apart at the anchor points. For safety, the crosser ties into the middle of a sling rope with a bowline around his waist. A figure-of-eight slip knot and snaplink on each end of the sling rope connects him to the top and bottom ropes. The snaplinks allow the safety rope to slide with the crosser as he progresses across the bridge.

- *Three-rope bridge.* (fig. 2—42). Details of construction of the three-rope bridge are fully described in TM 5—270.

Suspension Bridges and Cableways. Suspension bridges and cableways are described in detail in TM 5—270.

Figure 2—40 One-rope bridge.

Figure 2—41 Two-rope bridge.

Figure 2—42 *Three-rope bridge.*

Section V EVACUATION

2—19. Techniques of Evacuation

The techniques of evacuation are proven techniques. They are, however, all subject to improvement and should be discarded or modified as better methods of handling patients are developed.

When evacuating a patient from mountainous areas :

- Select the smoothest available route.
- Avoid unnecessary handling of the patient.
- Protect the patient from falling rock by placing his helmet over his head.
- When the evacuation route is long and arduous, a series of litter relay points or warming stations should be established. Warming stations, if established, must be staffed with the minimum medical personnel to permit proper emergency treatment of shock, hemorrhage, or other emergency condition.
- When a patient develops new or increased signs of shock while being evacuated, he should be treated and retained at one of these stations until such time as his condition warrants further evacuation.
- Skis can be used in various methods as aids to evacuation :

  Ambulatory patients can sometimes evacuate themselves on skis.

  Skis can be used as emergency splints.

  Several skis tied together can be used as a field expedient sled in lieu of the 200 lb sled (Ahkio).

  When covered with field jackets, skis make an expedient litter.

- Helicopters or heated vehicles should be utilized for evacuation whenever the situation warrants or the terrain permits.

2—20. Special Training

Prior to receiving training in basic mountain evacuation, litter teams should receive instruction in military mountaineering.

Litter bearers and aid men must be thoroughly familiar with the use and care of rope as an item of equipment.

The members of litter teams must become thoroughly proficient in the technique of belaying and in the choice of belay points so that proper support and protection can be given to patients and litter bearers when evacuating over steep difficult terrain.

2—21. Preparation for Evacuation

Three types of litters are available for evacuation of casualties in rough mountain terrain (fig. 2—43)—

- Standard collapsible canvas litter.
- Rigid Stokes litter.
- Rigid mountain basket litter.

The use of the rigid mountain litter and the Stokes litter will serve to eliminate the excessive knot-tying and rope lashing necessary to secure a patient to the standard collapsible canvas litter.

To prepare the standard collapsible canvas litter for evacuation in rough mountain terrain (fig. 2—44)—

Place sticks 40 cm long and at least 3 cm in diameter along each hinge brace and

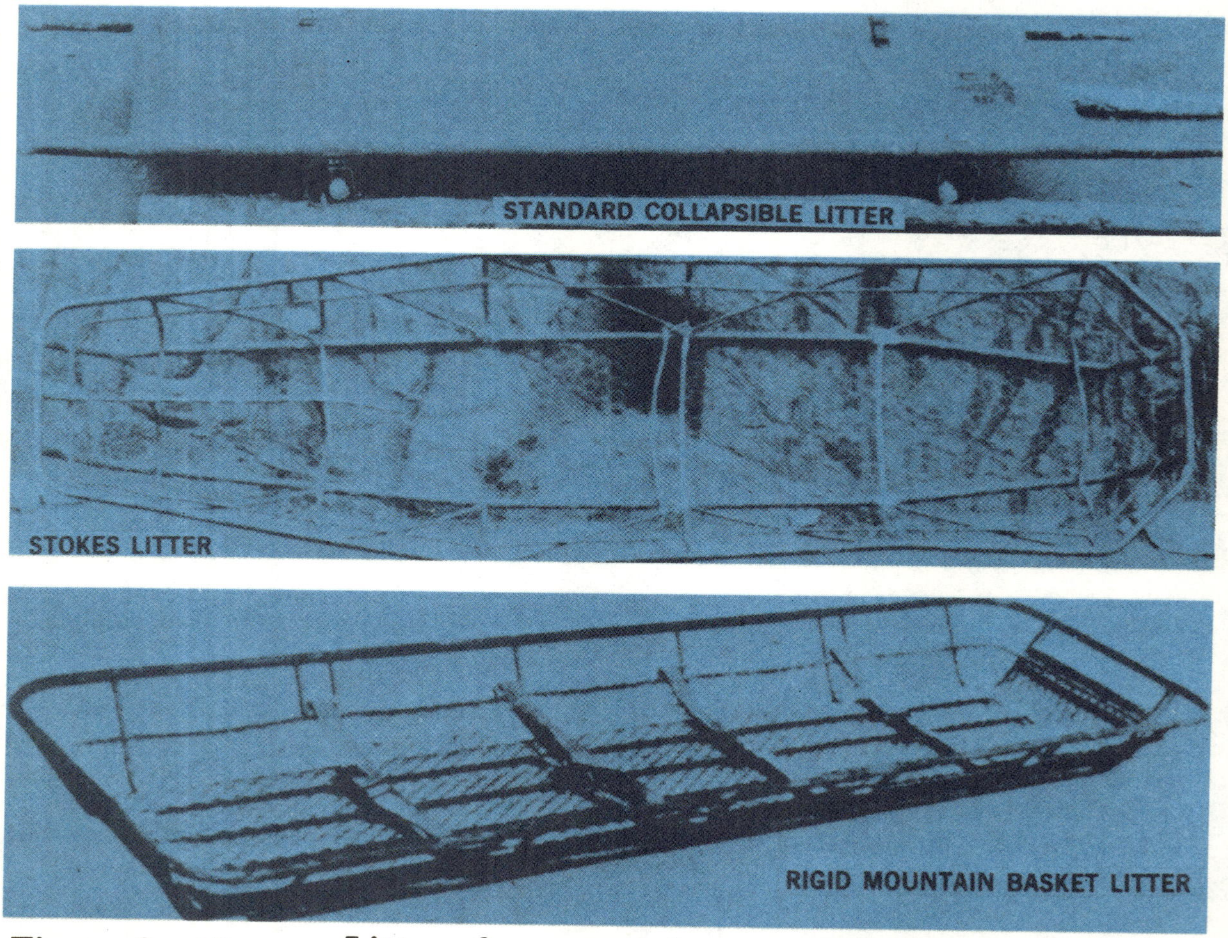

Figure 2—43 ***Litters for mountain evacuation.***

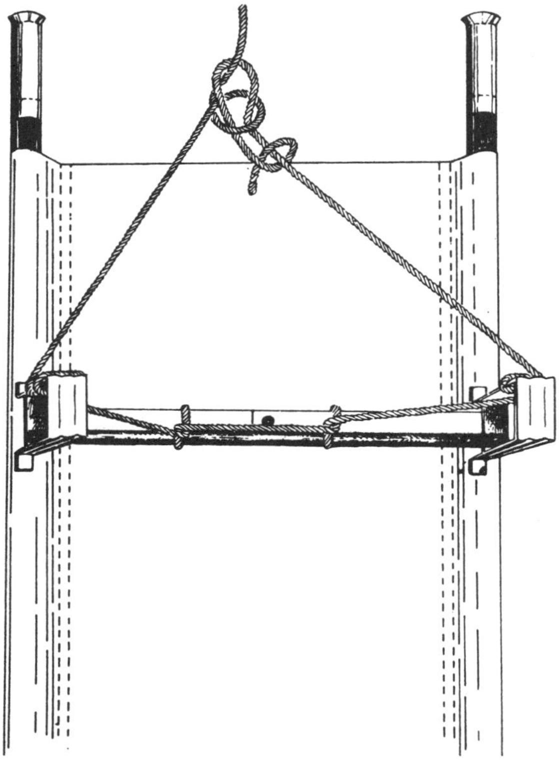

Figure 2—44 Method of fastening belay rope to litter.

fasten it securely to each stirrup with wire or strong cord. Lash the sticks to each brace along the length of the stick to prevent the hinge joint from collapsing.

Pass 2 or 3 meters of the end of a climbing rope through one litter stirrup, bringing it around and through again, forming a round turn.

Secure each 40 x 3 cm stick in place against the litter brace by two half hitches, one on each side of the hinge joint.

A round turn is then made on the opposite stirrup.

Tie the end to the long part of the climbing rope with a bowline between the canvas of the litter and the upper ends of the handles. Insure the knot is positioned so that it will clear the canvas of the litter.

The end thus prepared becomes the head of the litter.

The patient must always be lashed or strapped to the litter. This should be done under the supervision of a trained medic and only then after some determination of the extent of the injury has been made.

In securing a patient to the litter, the pelvic bones lend themselves most favorably to the tie since injuries of the leg, abdomen, chest or arms will not be involved in the procedure. A blanket is folded like a diaper over his abdomen. This eases the pressure caused by the sling at the points of contact. Four 4-meter sling ropes are used; two to lash the upper part of the body and two for the lower part. A team of two men is most efficient, one man working on each side of the litter. The upper part of the body is lashed first. Two sling ropes are used; they are tied first with a bowline on the upper part of the legs in the crotch, one sling rope on each leg.

The ends of the ropes are then brought diagonally across the body, under the arms, to the stirrups on the opposite sides of the litter, passed through the stirrup and a round turn formed; the rope is brought back across the chest and tied off with a square knot. If the ropes are not long enough to reach across the chest and back again, they should be tied off on the chest after the first round turn on the stirrup. To tie the lower part of the body, sling ropes are tied to each of the upper stirrups with a round turn and two half hitches, brought diagonally across the body to the stirrup on the opposite side of the litter at the foot, and round turns made on each stirrup. The knees of the patient are bent slightly and the feet drawn back. The ropes are brought from the stirrups and wrapped around the feet, bringing the rope across the bottom of the feet to act as a platform. Tie off with a square knot on the bottom side of the feet so that the knot will not form pressure on the arch of the foot. The knees are then pushed down tightening the ropes (fig. 2—45).

The arms of the patient may be tied separately to the litter when there are no arm injuries. In the event the patient panics or falls unconscious during evacuation, he would be unable to grab anything or cause a flopping arm or hand to become lodged.

In the event that the patient is wrapped in blankets or a sleeping bag, the above method cannot be used. A clove hitch is made with one loop about the small of the patient's back and the other loop about his hips, just below the buttocks. The running end from the upper loop is passed through the upper stirrups and across the patient's chest and secured. The running end from the lower loop is passed through the lower stirrups and secured over the patient's legs in a similar fashion.

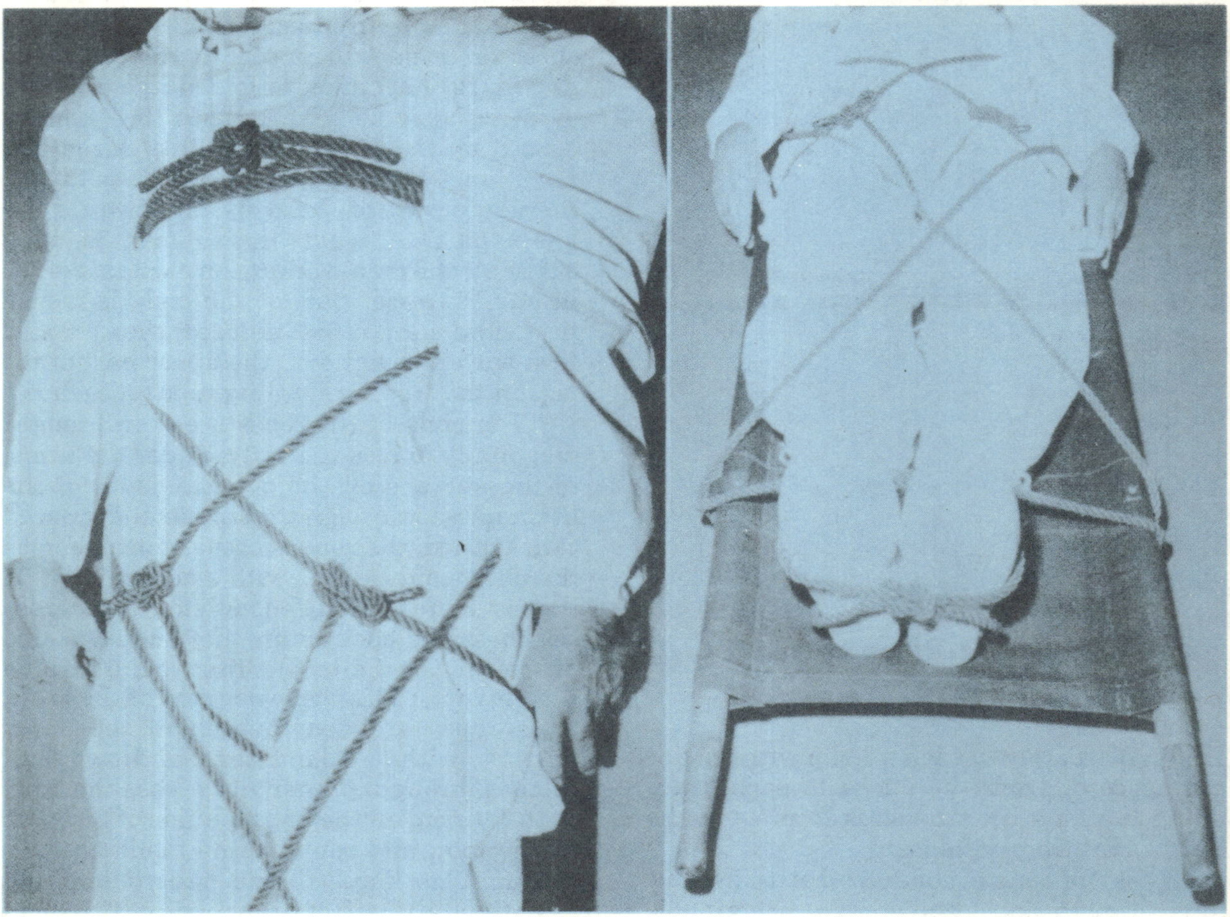

Figure 2—45 ***Securing casualty to litter.***

2—22. Ascending and Descending Slopes

Ascending. (fig. 2—46). In ascending a steep slope, one or two litter bearers take their places at the head of the litter, and a third, using the litter sling, takes his place at the foot. When two litter bearers are used at the head of the litter, a thin sapling is passed through the stirrups and extended 45 cm on either side of the lines affording a more secure grip for the two litter bearers at the head of the litter (fig. 2—47). The fourth and fifth men take their positions along the extended rope which is being held by the sixth, or belay man. On the signal "UP ROPE" the fourth, fifth, and sixth men pull while the others lift the litter and slowly climb. The men carrying the litter should not try to do all the work, but allow themselves to be partially pulled up the slope while they hold the litter off the ground. Positions are exchanged at each halt so that the work will be equally distributed. When each belay point is reached, the litter is placed on the ground and a new belay position farther up the slope is taken.

Descent. (fig. 2—47). In making a descent, one man acts as belay man and another takes his position on the rope, assisting him in lowering the litter. The three litter bearers

Figure 2—46 Steep slope evacuation—ascending.

take their positions as for the ascent. The sixth man may assist with the litter or precede the team, picking a trail, making the passage more negotiable by cleaning away shrubs and vines, and making a reconnaissance so that the team need not retrace its steps if a cliff should be encountered. In making the descent, the most direct and practicable passage should be taken. All available trees and rocks should be used as belay positions.

If the slope is relatively free of underbrush or trees, a litter travois is constructed and a three-man team used (fig. 2—48). In this procedure skids are placed under the litter. Two saplings, each 5 to 6 meters long and about 8 cm in diameter at the heavy end, are cut. The small ends are secured to the stirrups and litter brace at the foot of the litter; the butt ends are secured inside to the stirrups at the head of the litter. In this manner, the butt ends of the saplings project about 2½ to 4½ meters beyond the head of the litter and act as the foot of the litter. The second and third men are belay men, relieving each other as the rope is paid out, by assuming successive belay positions. In this way the descent need not be interrupted while the belay position is being changed. Using this method of evacuation, it has been demonstrated that a patient can be evacuated 60 to 70 vertical meters in about 10 minutes and that vertical faces of rock 1 to 2 meters high can be descended without difficulty.

2—23. Cliff Evacuation

Evacuation via cliffs should be used only when absolutely necessary and only by experienced personnel. The cliffs with the smoothest faces are chosen for the route. In this method, at least five standard climbing ropes are necessary.

The preferred method to be used requires only four litter bearers, five climbing ropes, and four 4-meter sling ropes. The litter is prepared by placing a stick at least 40 cm long and 3 cm in diameter along each hinge brace and fastening it securely to each stirrup with wire or strong cord. Two poles 2 to 3 meters long and about 10 cm in diameter are used as runners. The poles are attached to the stirrups by rope or wire; at the point of fixation, notches are made so that the rope or wire will not be worn unnecessarily by friction.

Two litter bearers remain at the top of the cliff, one to belay and the other to transmit signals and directions. The remaining two litter bearers each secure two rapelling ropes to suitable anchors at the top of the cliff. They each grasp one of the handles at the foot of the litter and working with the man on belay follows the signals and directions of the litter bearer who remained on the cliff, either playing out rope or holding the litter in position, the two rapellers guide the litter down the face of the cliff (fig. 2—49). Upon reaching the base of the cliff, the two rapellers carry the foot of the litter away from the cliff as the man on belay gently lowers the head of the litter until it is on the ground. This method can be worked very smoothly by a trained team with no added danger to the patient or the bearers.

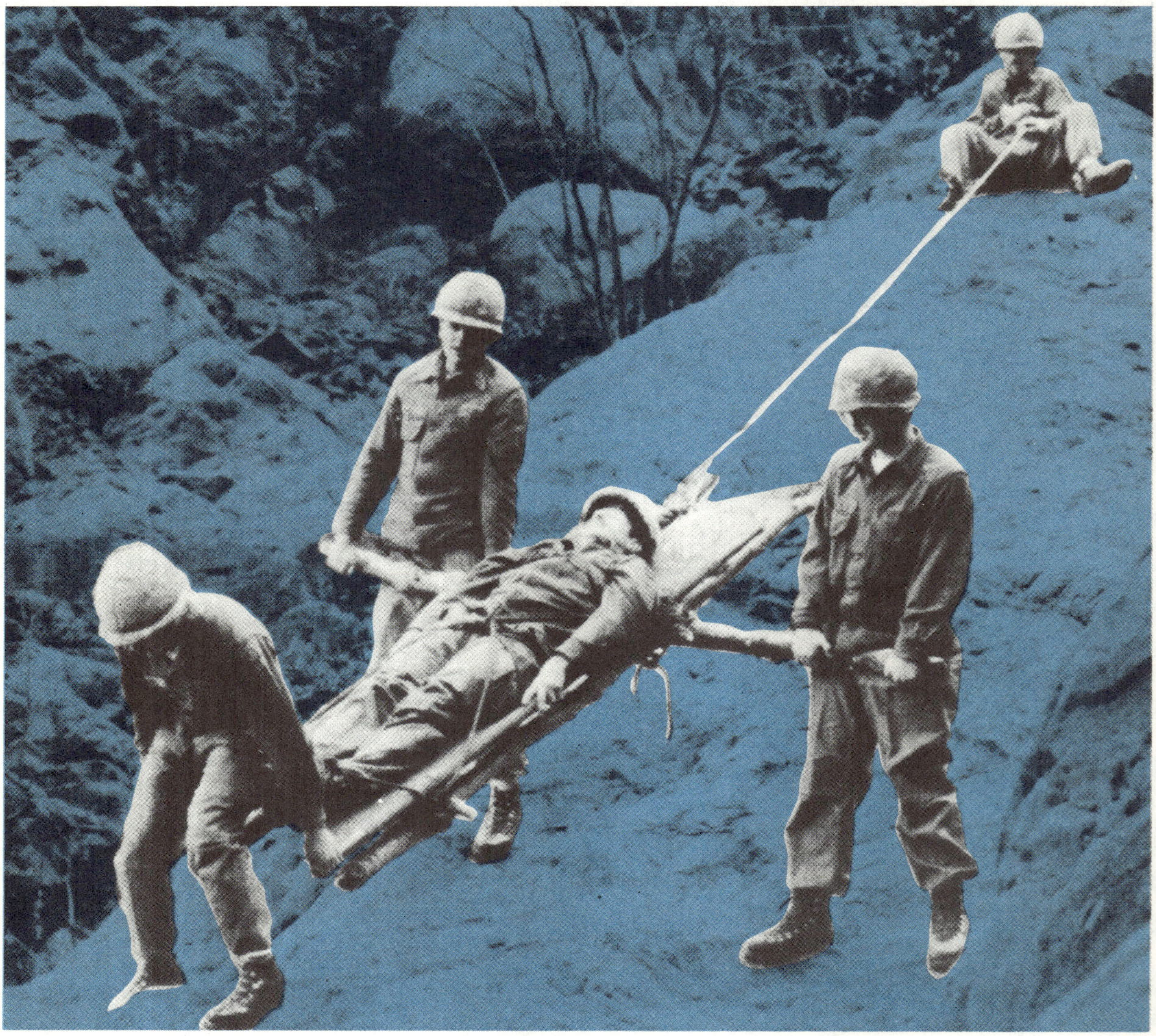

Figure 2—47 Steep slope evacuation—descending.

Figure 2—48 *Litter travois.*

The following is an alternate method: The patient is secured to the litter in the usual manner. Two poles, 2 meters long, are cut from trees with a diameter of 10 to 15 cm, and used as runners. Both ends of the poles are beveled. The poles are attached to the stirrups by rope or wire; at the points of fixation, notches are made so that the rope or wire will not be worn unnecessarily by friction. A guy rope is attached to each stirrup at the foot of the litter in order to prevent the litter from spinning during the descent. The litter is lowered over the cliff by two men, while a third man belays it. Meanwhile a fourth man escorts the litter on the descent by rappelling down a rope which parallels the route of evacuation, and assists the litter over projecting obstacles. The guy ropes are held taut by the fifth and sixth men who take widely separated positions at the bottom of the cliff. They can also help to ease the litter over protruding rocks by stepping away from the cliff and adding tension to the guy ropes. As the litter approaches the bottom of the cliff, the guy rope operators approach the litter, maintaining tension and, when close enough, grasp the litter handles and lower it to the ground.

The same procedure can be used in evacuating a patient up a cliff except for the following variations:

- An additional man will be needed at the top of the cliff to help raise the litter up and over the edge.
- Guy ropes must be attached to the stirrups at the head of the litter, instead of the foot.
- In the case of an overhanging cliff, an A-frame must be constructed at the top.

Figure 2—49 *Cliff evacuation—descent.*

2—24. Carriers

Patients not seriously injured but who cannot negotiate a descent by themselves may be carried down by a carrier who is belayed from above. The carrier, in making his descent, leans away from the slope as much as possible and walks sidewise allowing the belay man to support him. *Facing the slope and descending backwards should be avoided since it leads to unnecessary falls.* By keeping his lower leg well braced, keeping the weigh of the patient high on his back, and by grasping trees and rocks, he can descend very steep slopes with relative ease.

To secure the patient to a carrier with a rope, the carrier rests on his hands and knees while the patient straddles his back (fig. 2—50). A 4-meter length of nylon sling rope is used. The center of the rope is placed under the patient's buttocks. The right loose end is passed under the carrier's right armpit, across his chest, over his left shoulder, and under the patient's left arm; the left loose end is passed under the carrier's left armpit, across his chest, and over his right shoulder,

Figure 2—50 *One-man carry.*

and under the patient's arm, and the tie is made across the patient's back. If the carrier suffers from the pressure of the rope against his chest, the rope may be crossed behind his neck instead of over his chest, before securing the ends around the patient. The carrier's shoulders should be padded to prevent cutting by the rope.

An alternate method is to use two pistol belts hooked together and draped over the carrier's shoulders. The patient straddles the carrier and the belay man secures the loose ends of the pistol belts under the patient's buttocks. Slackness in the pistol belt sling should be avoided, since the carrier is most comfortable when the patient rests high on his back (see FM 8—35).

2—25. Field Expedient

The pole-carry method should only be considered as a last resort, since it is very difficult for the litter bearers to handle, and very uncomfortable for the patient. Its use is advisable only when narrow ledges must be traversed. The patient is wrapped in two blankets and suspended from a 4-meter long pole by a rope, pistol belts, or web belts. In suspending the patient, all excess slack should be avoided, since it increases the sidesway during carry. A two-man team, one on each end of the pole, carries the patient. If the pole is long enough, a four-man team may be used.

2—26. Evacuation From Trees

It is best to use two men for the evacuation of wounded or injured from trees, although one can accomplish the task by himself if necessary. Injured personnel in trees generally cannot help themselves, or they have fallen and become wedged in such a way that they are helpless. One man climbs the tree, taking one end of the rope with him; he passes the rope over a branch in the tree above the position of the patient and then ties a bowline on a bight. He slips one loop over each thigh of the patient and lowers him from the tree. The first man escorts the patient in his descent and prevents his movement from being impeded by intervening limbs or branches.

2—27. Horizontal Hauling Line

The horizontal hauling line (fig. 2—51) is one of the most complex of the various methods of evacuation discussed so far. It is primarily used in those cases where a steep slope or cliff has to be passed and where, at the same time, there is an intervening obstacle such as a swiftly running mountain stream. It can also be used to span a chasm when bridges are demolished. This method should be used only where there will be a considerable number of patients, such as at a warming station or collection point, and should not be installed for the evacuation of only one or two. It can also be used to lower or raise patients over obstacles.

The hauling line is installed as follows :

By means of a bowline, a 3 cm manila rope is secured to a tree far enough from the edge of the cliff (2 to 3 meters) to permit freedom of movement by the medical personnel. On the opposite side, the other end of the rope is passed around another fixed point (tree, boulder, or possibly a vehicle) and, by means of a transport knot, is pulled as taut as is necessary. All traverse ropes should have a certain amount of slack. When manila or sisal rope is used, a 5 percent sag should be allowed to avoid undue fatigue from developing in the rope.

To suspend the litter, two snaplinks are placed upon the traverse rope and one long litter carrying strap attached to each. A lower and upper retrieving rope is attached either to the litter stirrup or to the respective snaplinks. In the latter case, the loose ends of each rope are tied together above the center of the litter so that, when drawn up or down, both snaplinks move simultaneously.

After the patient has been secured to the litter, the litter is raised and the litter carrying straps or suspension ropes are passed through the stirrups and fastened together or else secured to the opposite stirrup.

To operate the hauling line :

- *Ascent.* Three men can easily raise the litter along the traverse by pulling on the upper retrieving rope. The pull should be steady and smooth in order to prevent jolting and sway.

- *Descent.* A gentle pull on the lower retrieving rope is enough to break the inertia and let gravity do the rest. During the descent, the men on the upper side should control the speed of the descent through their retrieving rope. It may be necessary to pull the patient the last few meters when the litter nears the low point of the slack in the traverse rope.

2—28. Expedients For Evacuation In Snow

Although litter bearers mounted on snowshoes will usually evacuate patients on snow covered mountains, downhill evacuation can be speeded up considerably when litter bearers are equipped with skis and a 200-pound capacity sled (Ahkio). For details of evacuation in snow, see FM 31—70.

Figure 2—51 Evacuation by horizontal hauling line.

Chapter 3

OPERATIONS ON GLACIERS and SNOW COVERED MOUNTAINS

Section I GENERAL

3—1. General Characteristics

Mountainous regions in cold and temperate zones are characterized during winter by—

- A great amount of snow.
- Extreme cold and severe snow storms which impose great hardships on troops.
- Snow falls at varying times in the autumn according to the geographical location, altitude, and exposure. The snow cover differs with the altitude and may also differ greatly from year to year. Deep snow cover greatly changes the problems of operations in the mountains, especially by increasing the difficulties of communication, transportation and sanitation. *Even a small amount of snow restricts the movement of men and vehicles.* The deeper the snow, the more it hampers and canalizes the movement of columns. If it is over 30 cm deep, movement of foot troops must be accomplished with the aid of skis or snowshoes. Wheeled vehicles will not be able to proceed any farther or faster than snow removal equipment will permit. The emplacement of artillery becomes an extremely slow and laborious process.
- Many slopes that are easy to negotiate in the summer become difficult and even dangerous to climb in the winter because of frozen and snow-covered ground and the risk of avalanches. If winds reach high velocity, traffic becomes almost impossible. Nevertheless, a great deal can be done to improve traffic conditions. If sufficient manpower and equipment are available, the principal routes can be kept clear. Snow removal becomes a continuous process because of drifting and frequent snowfalls.
- Snow conditions can occur in high mountains at all seasons of the year.

3—2. Effect of Special Training and Equipment

Small units of mountain troops, specially trained and equipped, can carry out operations in the mountains during the winter. Because of their training, they are aided, rather than retarded, by the snow. The importance and effectiveness of winter operations has taken on an added significance since the adoption of airmobile operations. Airmobility makes large scale winter operations in mountains entirely practicable.

398-868 O - 83 - 5

Section II CLASSIFICATION OF SNOW

3—3. Classification Of Snow

The terms *snow* and *ice* have been used in a general sense to describe snow or ice forms. In order to clarify the danger inherent in snow and ice, it is necessary to outline the forms that snow assumes in its transition to glacier ice. Some forms of snow are of considerable benefit to the mountain soldier, while others are a source of great danger.

Snow is classified as *powder snow* and *compact snow*. Powder is snow in its early stages and is further classified as *new snow, settling* or *settled* snow. Compact snow is snow which has passed beyond the stage of settled snow.

New snow maintains individuality of flakes immediately following its fall. If new snow is dry, it is feathery; if damp, it quickly consolidates into settling or settled snow.

In areas of extreme cold and at high altitudes, new snow not uncommonly comprises two forms:

- *Sand snow* falls at extremely low temperatures. As its name implies, it is sandy. Skis and sleds slide over it with difficulty.

- *Wild snow* is very dry new snow which falls during complete calms and low temperatures. It is extremely light and feathery and easily lifted and moved by wind.

New firn snow is compact and granular and is composed of closely associated grains instead of individual flakes. Heat-softened new firn snow is referred to by skiers as *spring or corn snow.*

Advanced firn snow or *neve* consists of closely associated grains of ice, separated by air spaces.

Glacier ice is composed of crystals of advanced firn snow cemented by a film of ice. There are no air spaces between the grains and any air spaces present are within the grains themselves.

3—4. Snow Formations

The forms of snow described above are subject to various modification during their transition from dry new snow to glacier ice. Temperature, humidity, and wind are important modifying factors in this transition. The following phenomena are important to the mountain soldier:

Sun crust. Any snow that has had the superficial layer melted by heat and subsequently refrozen. A layer of snow that is sun crusted and weathered throughout its thickness becomes firn snow. Sun crust commonly overlies powder snow.

Wind crust or windpacked snow. Usually found on windward slopes and is anchored firmly to the underlying snow. A moist wind blowing over snow slopes causes the surface snow to become compacted. Depositing of drifting snow is not involved. Inasmuch as the windpacked snow has not been transported, the crystals are not greatly affected by abrasion. Consequently, they reflect much light. Wind crust is generally safe.

Wind slab. Formed from snow transported and deposited by winds. While the slabs themselves are well compacted, they are loosely anchored to the underlying surface and generally will have an air space between the slab and the underlying snow. Having been transported by wind, the component grains are rounded and do not reflect the light. For this reason the surface of a wind slab has a dull, chalky appearance. *Wind slabs are extremely dangerous* because they often form the platform for subsequent avalanches that may develop.

Section III AVALANCHES

3—5. General

It is not known for sure when an army first suffered casualties from avalanches although it is possible that there were some during Xenophon's Retreat of the Ten Thousand in 401 B.C., or during Alexander The Great's Asian Campaigns around 330 B.C. It can be assumed that Hannibal suffered thousands of avalanche casualties during his crossing of the Alps in 218 B.C. Avalanches again played an important part in 1800 when Napoleon waged his war in the Alps. Overwhelming military disasters occurred during World War I when at least 40,000 men died in avalances in the Tyrol and South Tyrol. It has been estimated that during fighting on the Austro-Italian front in 1916, as many as 9,000 to 10,000 troops were killed in a single avalanche period of two days duration.

The danger of avalanches must be carefully considered when military operations are being conducted in snow covered mountains. The danger of avalanches has become greater in modern warfare because of the vibrating effects caused by nuclear detonations. These effects may cause severe snow and rock slides many kilometers away from ground zero.

3—6. Basic Causes of Avalanches

General. Fundamentally there are only two causes of avalanches—*large amounts of snow* and *steep slopes for it to slide on.* These two essentials, acting together, can produce an avalanche without any other assistance.

Terrain. Ground surface conditions have considerable effect upon snow in contact. A broken, serrated, or boulder-strewn surface provides a good anchorage for a snowpack.

Figure 3—1 *Avalanche slopes.*

Slides breaking off at ground level are unlikely. Smooth, even slopes of bare earth, solid rock, or shale favor massive ground-level avalanches, typical of the high alpine zone. Contours of a mountain influence the avalanche; terraces, talus, basins, and outcrops are effective barriers. They either divert the moving snow or give it room to spread out and lose its momentum. Gullies collect and channelize the descending snow, making favorable slidepaths which must be carefully avoided (fig. 3—1). Ridges lying parallel to the slidepath, on the other hand, are relatively secure. The convex slope is more likely to avalanche because the snow layers settling upon it are placed under tension. Avalanches usually fracture at the sharpest point on the curve, get up to full speed instantly, and pulverize rapidly. The steepest part of a convex slope is generally near the bottom, leading to a sudden transition and poor anchorage at the toe of the snow layer. On a concave surface, snow is under compression. The steepest part of the slope is generally near the top; the transition is more gradual and the anchorage at the toe is better than for a convex slope. The dimensions of the slope (length and width), are other factors that determine the size of the snow slide and possible destruction caused.

All of the terrain features mentioned above are less important than grade. The steeper the slope, the more likely it is to slide regardless of any other conditions. The minimum angle favorable to avalanches is 25°. Slopes from 25° to 35° may avalanche, especially if disturbed by cutting action of a skier or some other factor. *The critical zone lies above 35°*. From this angle to the point where snow can no longer cling, except by windpacking, slopes are apt to slide in some fashion with every storm.

Vegetation. Vegetation of any kind, *except grass,* has a restraining effect on avalanches. The existance of heavy forest cover is an indication that slides in the location are rare or of minor importance. It is a mistake, however, to consider all forested areas as safe. Scattered timber is no particular deterrent. Slopes where the timber has been destroyed by fire offer good potential for snow slides.

Exposure. Slopes facing the sun favor avalanches produced by thawing. Loose snow avalanches are more common on slopes opposite the sun. Cornices habitually form along ridges and crests lying at right angles to the prevailing wind. Lee slopes are the most probable locations for overloads of wind-driven snow and formation of slab. On the other hand, snow is transported from the wind-beaten slopes and that which remains is packed and stabilized.

Climate and Weather. In addition to the terrain factors, climate and weather are the other basic ingredients for the avalanche phenomenon.

Storms which deposit up to 2.5 centimeters of new snow per hour are not uncommon.

Temperature fluctuates widely and rapidly in the mountains. Prolonged spells of extremely low temperatures occur and there are occasional intrusions of warm air masses, usually in connection with a storm. Rainfall may occur in the coastal zones and create avalanche conditions. Temperature greatly affects the cohesion of snow; a rise in temperature retards settlement of the snow mass and increases the brittleness and tension of a slab formation.

Wind action during storms in the mountains is generally strong and its influence on

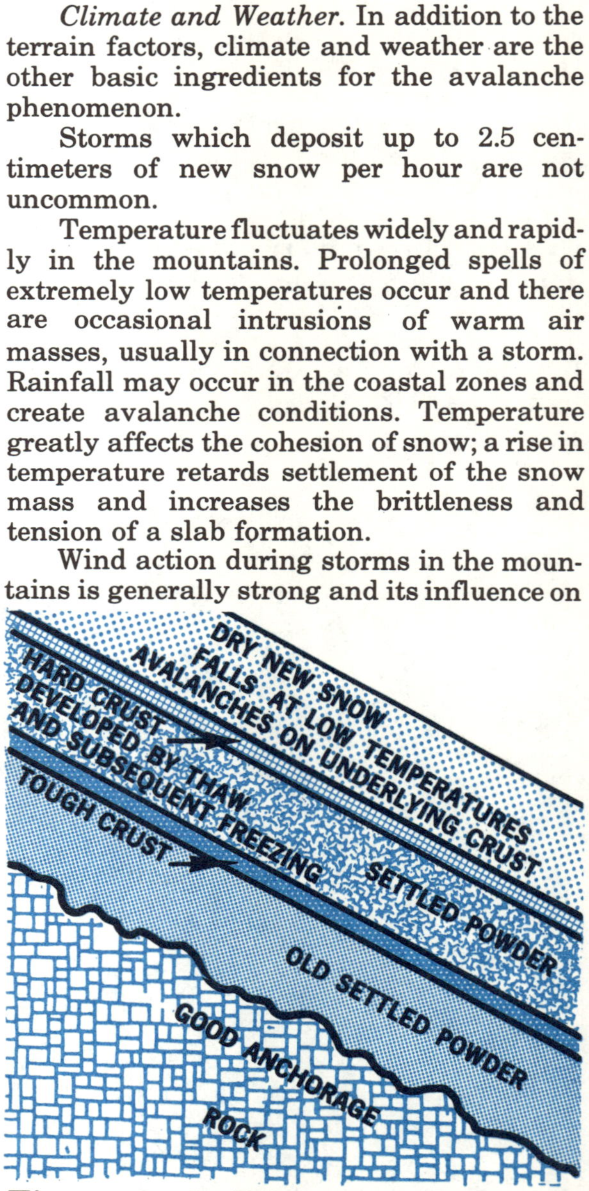

Figure 3—2 Dry snow avalanche —composition of snow layers.

Figure 3—3 Dry, loose snow avalanche.

snow is the most important of all the contributory factors. It transports snow from one exposure to another during storms and fair weather, thus promoting overloads on certain slopes. It also modifies the size and shape of snow particles.

3—7. Types of Avalanches

Classification. Avalanches may be classified in accordance with the type of snow involved, the manner of release, or the size. For practical purposes, classification according to the type of snow involved is normally used and all slides divided into the following two general groups:

- Loose snow avalanches.
- Packed snow avalanches.

Loose Snow Avalanches. Characteristics. An avalanche of loose snow always starts on the surface from a point or a narrow sector. From the starting point it grows fanwise, expanding in width and depth. The speed and nature of its development depend on whether the snow is dry, damp, or wet.

Dry loose snow avalanches. These are composed of loose snow, possibly drifted but not windpacked. They normally start at a point of origin and travel at high speed on a gradually widening path, increasing in size as they descend (figs. 3—2 and 3—3; note the enormous cloud of snow dust involved). A dry loose snow avalanche is always shallow at the start and depends on the snow it can pick up during its run for volume. Thus, a dry snow avalanche of dangerous size can only occur on a long slidepath, or from a large accumulation zone which funnels into a constricted outrun. Occasionally, heavy snowfall at low temperatures produces the phenomenon of the "Wild Snow" avalanche—formless masses pouring down the mountainside. They are actually avalanches of air and snow mixed. Windblast, a side effect of very large, high speed avalanches, is powerful enough to damage structures and endanger life outside the actual avalanche path. The hazard from loose snow avalanches is soon over.

Damp and wet snow avalanches. These resemble those of dry snow with the same

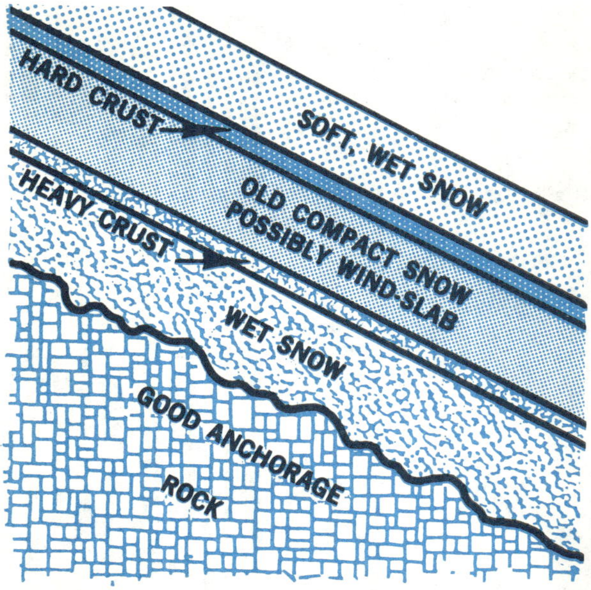

Figure 3—4 Wet snow avalanche —composition of snow layers.

arrowhead point of origin, gradually becoming wider. Their mass is many times greater than that of a dry avalanche and they are therefore much more destructive. Being heavier and stickier, however, they develop more friction and therefore travel at a slower pace. The principal hazard of damp and wet snow avalanches is to fixed installations. Their comparatively low speed causes them to stop rather suddenly when they lose momentum and to pile up in towering masses. This is in contrast to the dry slide which tends to spread out like the splash of a wave.

Damp and wet slides (fig. 3—4 and 3—5) solidify immediately upon release from the pressure of motion; adding to the difficulty of rescue or clearing operations. Wet slides have a distinctive characteristic of channeling. The moving snow constructs its own banks and flows between them like a river of slush, often in unexpected directions. The damp sunslides of midwinter are generally shallow. But the wet avalanches of spring, caused by deep thawing either from rain or prolonged temperatures above freezing, often involve enormous masses of snow and debris and have tremendous destructive power.

Packed Snow Avalanches.
Windslab

Windpacked snow, called windslab or snowslab is unquestionably the worst killer of all and equal to the wet spring avalanche as a destroyer of property. Hard slab is usually the result of wind action on snow picked up from the surface. Soft slab is usually the result of wind action on falling snow. Windslab avalanches behave in an altogether different manner from loose snow. It has the ability to retain its unstable

Figure 3—5 Wet snow avalanche.

character for days, weeks, and even months; thus leading to the unexpected release of delayed-action avalanches, often from minor causes such as the addition of a small amount of new snow, a skier cutting across a hillside, sun action, or even a minor vibration.

The windslab avalanche combines great mass with high speed to produce maximum energy. It may originate either at the surface or through the collapse of a stratum deep within the snowpack. It starts on a wide front with penetration in depth. The entire slab field—top, sides, and bottom—releases almost simultaneously. The place where the slab has broken away from the snow pack is always marked by an angular fracture line (fig. 3—6; note that it broke at the sharpest part of the curve on a convex slope) instead of a point, roughly following the contour. In a packed-snow avalanche the main body of the slide reaches its maximum speed within seconds. Speeds of 100 kilometers per hour are not uncommon. Thus it exerts full destructive power from the place where it starts, whereas a loose-snow avalance does not attain its greatest momentum until near the end of its run.

Due to the characteristic delayed release action as stated above, the *slab avalanche is the most dangerous of all types.* A series of slab avalanches may

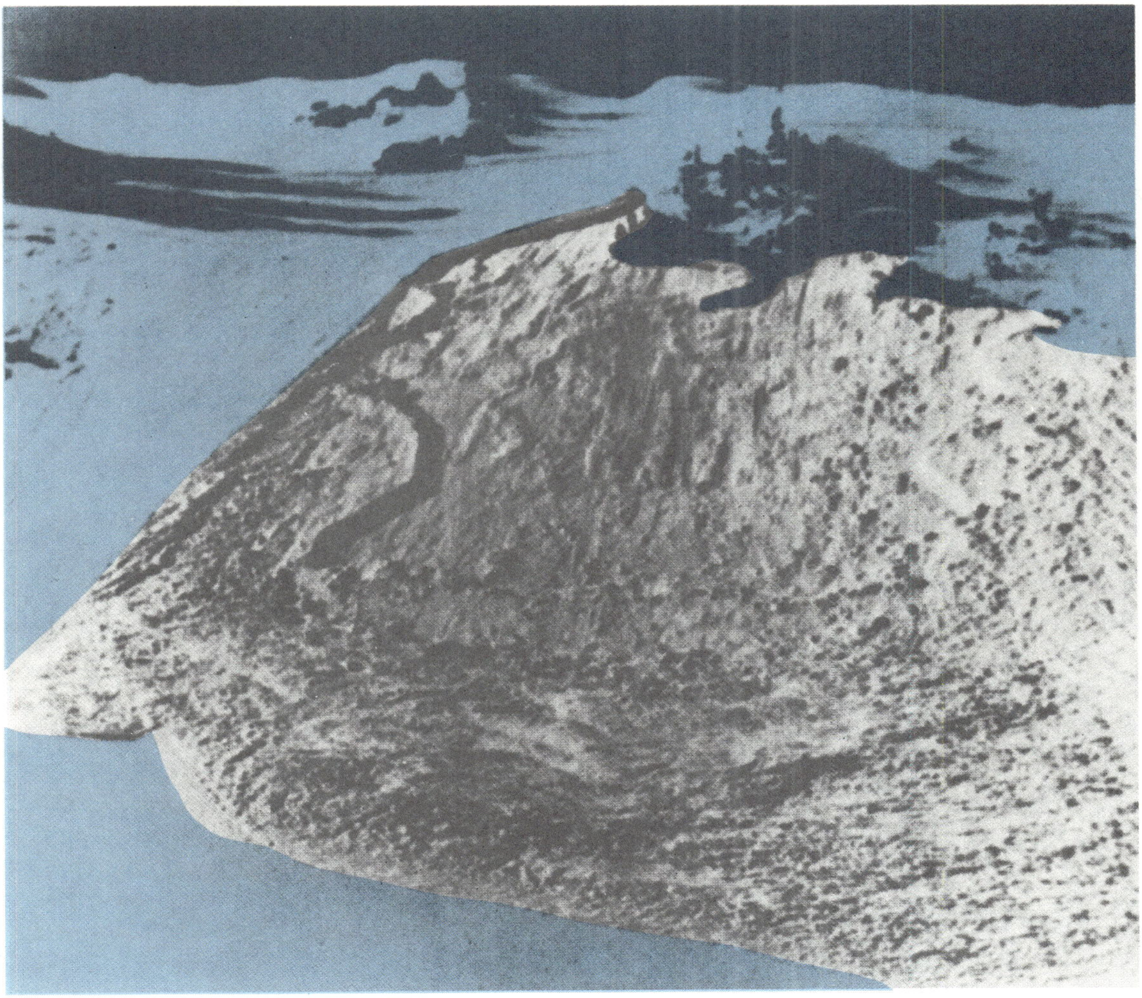

Figure 3—6 ***Slab fracture.***

stabilize conditions only locally, leaving an adjacent slab as lethal as an unexploded shell. If found on the surface, windslab has a dull, chalky, non-reflecting appearance and has a hollow sound underfoot. If the slab is hard, it often settles with a crunching sound which an experienced mountaineer recognizes as a danger signal. The only way to insure that movement across slab formation is safe, is to break the slab by one of the stabilization methods described in paragraph 3—10.

Cornices. A cornice (fig. 3—7) is a snow formation allied to the slab. They build up on the lee side of crests and ridges which lie at or near right angles to the wind. Occasionally they are straight-walled, but their characteristic shape is that of a breaking wave (fig. 3—8). The obvious hazard from cornices is due to the fractures of the overhang from simple overloading, weakening due to temperature, rain, or sun erosion. These falling blocks are generally large enough to be dangerous by themselves. They may also release avalanches on the slopes below.

Combination Avalanches.

General. Combination avalanches are composed of both loose and packed snow formations making it more difficult to determine which part of the combination acted as the trigger and which is the main charge of the avalanche.

The climax avalanche. The climax avalanche is a special combination type. The distinguishing characteristic of this type avalanche is that it contains a large proportion of old snow and is caused by conditions which have developed over a considerable

Figure 3—7 **Cornice.**

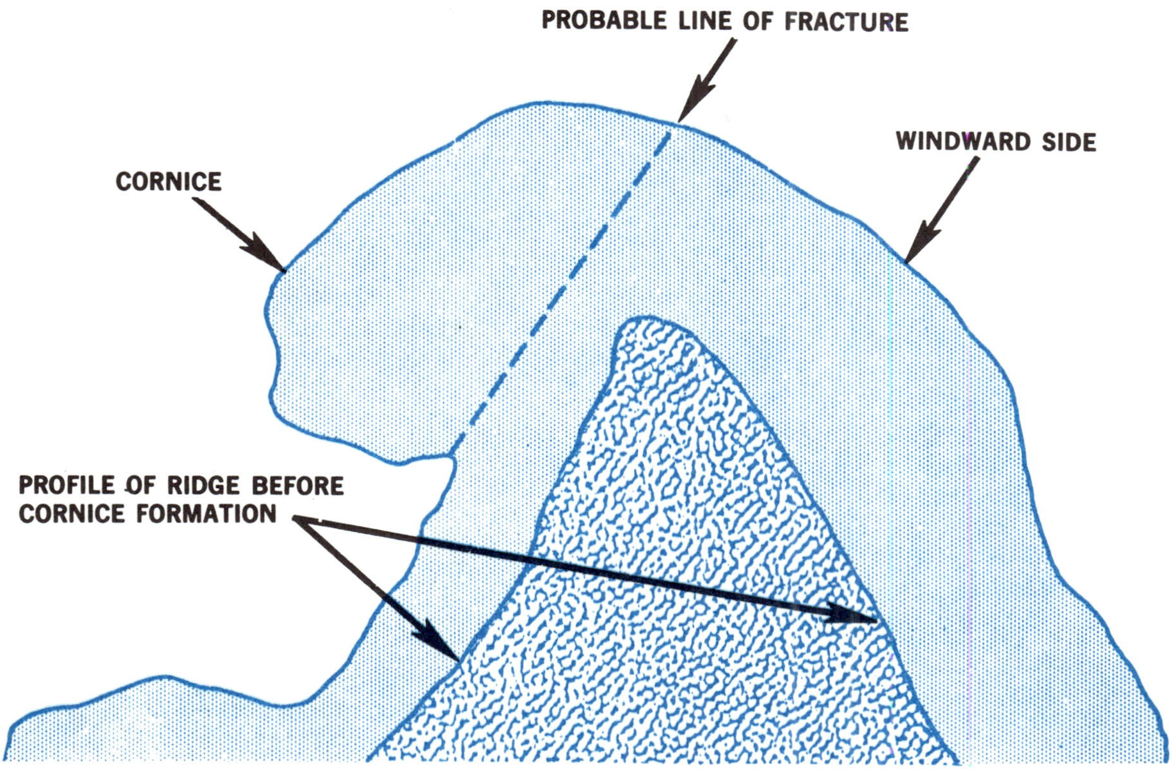

Figure 3—8 *Composition of a cornice.*

period of time—at least a month and possibly an entire season. Climax avalanches occur infrequently because they require an unusual combination of favorable factors. Whenever they occur, the penetration of a climax fracture is always in great depth, usually to the ground. They travel farther and spread out wider than ordinary avalanches on the same slide path.

3—8. Avalanche Triggers

Triggering. A loose-snow slide usually occurs during or immediately after a storm or some other weather situation that creates instability. A slab avalanche may come as a delayed action. In any case, every avalanche must have a "trigger." There has to be some final nudge, some force or combination of forces, to account for the release of masses of snow at a particular time and place.

Classification. Four avalanche triggers are recognized:

- Overloading.
- Shearing.
- Temperature.
- Vibration.

Overloading. Weight is probably the most frequent cause of avalanches. New snow piles up until it overcomes cohesion and the structure collapses of its own weight and begins to slide.

Shearing. This can be applied in various ways—the slicing action of a pair of skis, a wad of snow falling out of trees or over a cliff, or any exterior force that cuts the bond. The vibration of a slide in motion has a shearing effect on the snow beneath.

Temperature. Temperature triggers avalanches by its effect on the cohesion of snow. A rise in temperature weakens the bonds, while a fall in temperature retards settlement of the snow mass and increases the brittleness and tension of a slab.

Vibration. This factor is related to shearing, but it is treated separately. Unlike the other triggers, it can operate at long range. Avalanches may be released by thunder, shock waves from high speed aircraft, explosions, earthquakes, and by other loud or sharp sounds—vibrations transmitted through the earth and snow by the movement of tanks, bulldozers, or heavy machinery. They can also be triggered by the primary or reflected blast waves resulting from the detonation of nuclear weapons.

3—9. Avalanche Hazard Forecasting

General. The accurate prediction of an avalanche occurrence either in time or location is impossible. The actual avalanche occurrence is governed by a set of variable factors which cannot be reduced to a simple mathemathical formula. The experienced mountaineer can usually recognize the development of a hazardous situation in sufficient time to avoid the danger area.

Terrain and Climate. Terrain and climate are the two basic causes of avalanches. Terrain is a constant, always prepared to do its part. Climate is the variable.

The Contributory Factors. Systematic avalanche studies conducted in the past have identified 10 factors which contribute to the avalanche hazard. These factors which are subdivisions of climate and weather are as follows:

- *Depth and condition of the base.* 60 cm depth is generally sufficient to cover ground obstructions and provide a smooth sliding base. Greater depths will obliterate such major natural barriers as terraces, gullies, outcrops, and clumps of small trees. If the bottom snowlayer consists of granular snow (also referred to as depth hoar), the slope is dangerous because the snowpack has no anchorage at its underside. Depth hoar is usually at the bottom of the snowpack, and can be detected by reversed ski poles, skis, probes, or vertical trenching.
- *Old snow surface.* A loose snow surface promotes good cohesion with a fresh fall but allows deeper penetration of any avalanche that starts. A crusted or windpacked surface means poor cohesion with the new snow but restricts the avalanche to the new layer.
- *Depth of new snow.* 30 cm of new snow is regarded as the minimum requirement to produce, by itself, an avalanche of dangerous proportions.
- *Types of new snow (especially its free moisture content).* Very dry snow has few cohesion qualities and will avalanche so readily that it seldom builds up to dangerous volume except under true blizzard conditions. The moisture in damp snow acts as a cement and improves cohesion. Wet snow is saturated. The water content "lubricates" rather than cements. Transition snow (on the dividing line between dry and damp) is especially susceptible to the formation of slab under wind action.
- *Average density.* The average density (water content per centimeter of snow) of dry snow varies from .15 to .25 cm of water per 2.5 cm of new snow. Very dry snow, typical of the high alpine zones but found occasionally in other areas, is much lighter; densities as low as .10 cm are common. Damp and wet snow, in contrast, has a density as high as .35 cm. These are normal densities and have little significance. When dry snow density exceeds 10%, its weight may be increasing faster than its cohesion and a slide may occur.
- *Snowfall intensity.* When the snow piles up at the rate of 2.5 cm or more per hour, the pack is growing faster than stabilizing forces, such as settlement, can take care of it. This sudden increase in load may fracture a slab beneath and result in a slide.
- *Precipitation intensity.* Based upon experience, it is concluded that with a continuous precipitation intensity of .25 cm of water or more per hour and with wind action at effective levels, the avalanche hazard becomes critical when the total water precipitation reaches 2.54 cm. This is one of the

newest methods used in avalanche forecasting. It requires interpretation by trained weather station personnel with special equipment.

- *Settlement.* Settlement of snow is continuous. With one exception, it is always a stabilizing factor. The exception is shrinkage of a loose snow layer away from a slab thus depriving it of support. In new snow a settlement ratio less than 15% indicates that little consolidation is taking place; above 30%, stabilization is proceeding rapidly. Over a long period ordinary snow layers shrink up to 90%, but slab layers may shrink no more than 60%. Abnormally low shrinkage in a layer indicates that a slab is forming.
- *Wind action.* Wind action is an important contributory factor. It overloads certain slopes at the expense of the others, it grinds snow crystals to simpler and less cohesive forms, it constructs stable crust and fragile slab, often side by side. Warm wind ("chinook" of North America and the "foehn" of Europe) is an effective thawing agent, even more effective than sunlight. By sudden changes in direction and velocity, wind can act as a shearing trigger on a layer of snow it has just deposited. An average velocity of 15 knots is the minimum effective level for wind's action in building avalanche hazards.
- *Temperatures.* Temperature directly influences the snow type. Dry snow normally falls at 25° F. and below. Temperatures above 28° F. promote rapid settlement and metamorphosis of the snow is sometimes too rapid. A sudden rise of temperature causes a loss of cohesion fast enough to trigger an avalanche. A sudden drop increases tension particularly in slab. Gradual warming in spring leads to cumulative deterioration of the snow and to heavy, wet avalanches.

Avalanche Conditions. Avalanches obey mechanical laws which can only be identified and evaluated by trained personnel with special equipment. The soldier in the field must generally rely on the following factors:

- Terrain—grade 20° or steeper.
- Old snow depth—enough to cover ground obstructions.
- Surface crusted—normally only new snow will slide.
- Surface loose—good cohesion between layers but both old and new snow may slide.
- New snow depth—25 cm or more.
- Snowfall intensity—2.5 cm per hour or more. This can be assumed when snowfall is heavy enough to restrict visibility to 100—200 meters.
- Precipitation intensity—.25 cm per hour of water or more plus strong wind action. This can be assumed if snowfall intensity is 2.5 cm per hour and snow is damp or noticeably heavy for dry snow or dry snow is granular or pellet in type.
- Settlement—noticeably low. Watch the snow collars around the trees or posts.
- Wind— 15 knots or higher average. This can be assumed if snow is blown parallel or almost parallel to ground.
- Temperature—any sudden change up or down. For example, a thawing temperature day and night for 36 hours; that is, no overnight freeze.
- Depth hoar—must be assumed on any avalanche slope in the high alpine zone.

3—10. Protective Measures

Several methods can be used to reduce the effects of the avalanche hazard. Some of the protective measures that may be taken by units in avalanche areas are—

Restrictions. Based upon terrain analysis and reconnaissance, areas which are considered hazardous may be placed "off limits" for all troops. This may affect only a few narrow avalanche paths, or an entire valley, or several valleys depending upon the terrain and weather conditions in the area of operations. Enforcement of restrictions may be a necessity even if the troops involved are required to conduct long and time-consuming detours.

Stabilization. In combat, individuals and units may be required to take calculated risks and enter areas described as hazardous. The risks can be considerably reduced by application of the following stabilization methods:

- *Stabilization by skiing.* By this means, the snow is caused to settle on the dangerous slope. Constant use of the hazardous slidepath area also prevents the snow from building up avalanche conditions. The work is done by teams of expert skiers (2 and 3 on each team). Great care, coordination between teams, and supervision must be exercised because of the dangerous nature of the work. The leading man in the party, using avalanche cord and being belayed by other members of the team, skis the release points of the slopes. As a result of this skiing action the snow will be either stabilized or the slide triggered; in either case the avalanche hazard is removed.
- *Stabilization by means of explosives.* Under extremely dangerous conditions it may be safer to stabilize the snow by using hand-placed charges or hand grenades. The safest way to do this is to throw charges of approximately 5 pounds over a crest, having the electric wire coiled so that the charge lands on the slope 5 to 6 meters below the crest. The charge may then be safely detonated from behind the crest. Under some snow conditions more powerful charges may be required. As a rule of thumb, 5—10 pounds of TNT controls 40 meters of slope width. Huge cornices are blasted by digging charges into the snow along the probable line of fracture. Individuals digging holes and placing charges must be belayed while working.
- *Stabilization by artillery and infantry weapons.* Carrying explosives and demolition equipment up mountain slopes in deep snow is hard work. An easier way to cause the snow to slide is to use artillery pieces within their range. Due to the difficulty of moving artillery pieces off the road or over secondary mountain trails, recoilless rifles are the handiest tool in campaigns against avalanches.
- *Stabilization by aircraft.* The use of artillery pieces as well as recoilless rifles is limited by their range. Avalanches may also be set off against lines of communications deep in enemy territory by use of aircraft. The pilot selects suitable slopes and causes the snow to slide by using guns, rockets, or sonic boom.

Use of Barriers. Lines of communications and fixed installations which are under avalanche threat can be protected by construction of avalanche barriers. Barriers can be formed by adding rocks and earth, concrete, or other similar materials to natural obstacles. This barrier will absorb some of the tremendous energy of the slide. Diversion walls and piers may also be used to divert the avalanche to a predetermined path or turn it away from the area to be protected. Because of the amount of effort involved, such work is normally undertaken only during static situations.

3—11. Safety Rules

General. Many of the hazards peculiar to avalanche areas can be avoided or their danger greatly reduced by knowing and practicing a few simple, common sense rules. The following rules should be adhered to whenever it becomes necessary to negotiate potential avalanche areas.

Adjusting Equipment. To insure freedom of legs and arms, the ice axe safety strap is removed from the wrist. When using skis, the bindings are loosened to facilitate their removal and ski pole straps are removed from the wrists. Skis tend to get tangled, restrict movement, and get dragged under the sliding snow. When snow conditions permit, such danger zones may be negotiated more safely on foot, carrying the skis. While skiing, all movements are executed with caution. Packs and weapons are either removed or loosened.

Selection of Routes. Learn the avalanche paths and, whenever possible, detour around the hazardous slopes. The crest of a ridge is a safe route, but never move on the overhang of a cornice. Avoid moving along the bottom of narrow, V-shaped valleys (point X fig. 3—9) where a disturbance could cause a double avalanche and fill the narrow valley floor with masses of snow almost instantaneously. Movement along the middle of wide, U-shaped valleys is less dangerous, since avalanches occurring there would tend to expend their force on the wide floor of the valley with little danger to troops moving in the middle. Do not move above natural barriers such as moraine (point Y fig. 3—10) located on the valley floor, since a slide could collide with the barrier.

Weather. Obtain the latest weather forecast from the best available source prior to starting a move. During the move be alert for any signs of change in the weather. Any sudden change can be dangerous.

Ascent and Descent. When avalanche slopes must be ascended or descended they are climbed as straight up or as straight down as possible, preferably along a line of protruding rocks, ledges, and trees where the snow cover is less likely to slide. Traversing back and forth is avoided since the snow cover is dissected and greatly disturbed. Ravines, gullies, and low terrain features which are locations of snow deposits are avoided as much as possible.

Observation and Testing. The area to be traversed must be constantly observed for signs of danger. It should be tested whenever the tactical situation will permit. Before entering a dangerous slope, a small area of the same grade and exposure should be tested. Windslab formations should be avoided, since they give a false impression of their solidity. Watch the shadows; when they point at a slope like warning fingers, sun action is at its maximum. Seek the protection of heavy timber, windbeaten slopes, and terrain barriers.

Proper Timing. In movement over areas of avalanche danger, distance between individuals is greatly increased and main-

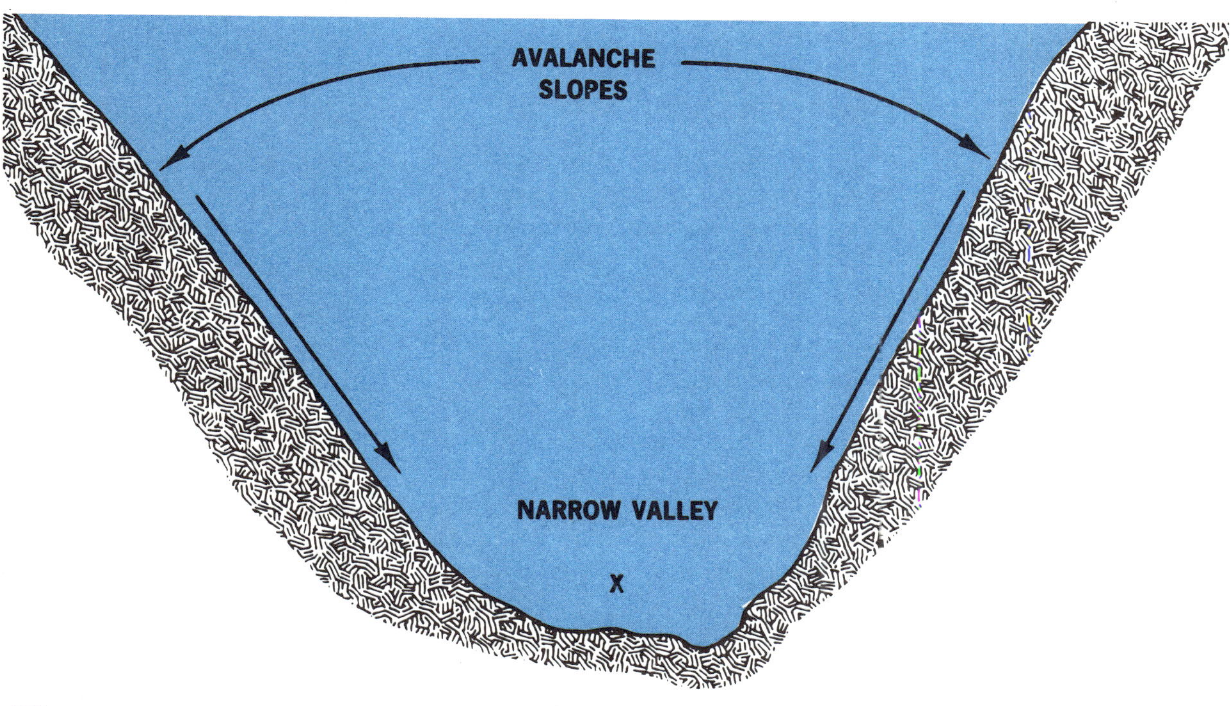

Figure 3—9 ***Danger of double avalanche.***

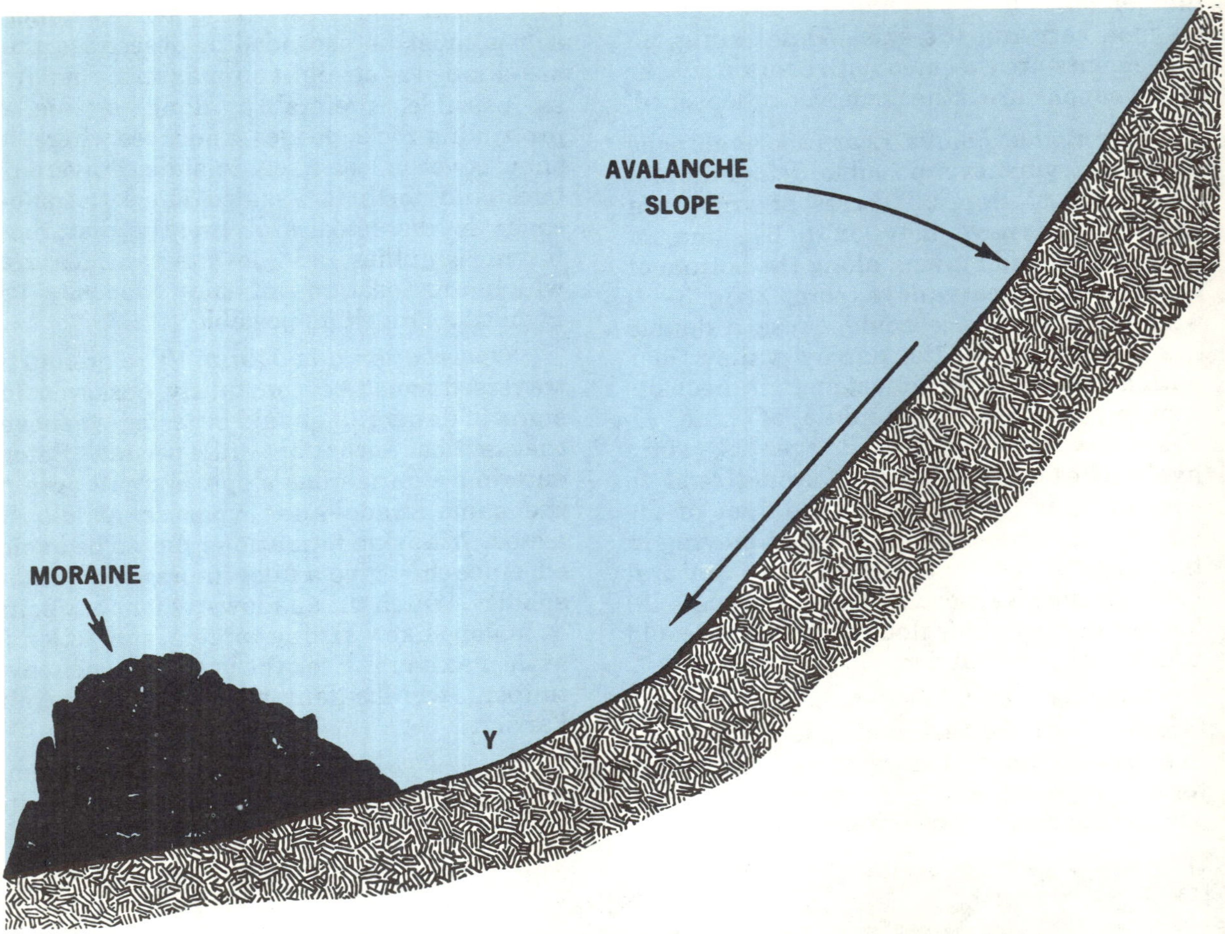

Figure 3—10 ***Danger of avalanche against moraine barrier.***

tained in accordance with prevailing conditions and the area to be negotiated. Only one or the least possible number of individuals are exposed to this danger at one time. If a dangerous slope, deep hollow, or gully must be traversed, let one man test the slope, belayed with the climbing rope by other members of the party. Visual contact is constantly maintained and, if necessary, warning posts are established or observers are stationed at safe locations along the route.

Use of Avalanche Cord. On reconnaissance missions in avalanche areas and especially when testing, one end of a brightly colored cord (15 to 20 meters in length) is tied around the body and trailed behind the individual. This brightly colored avalanche cord is an additional safety measure and facilitates location and rescue of individuals buried by an avalanche, since the cord tends to remain on or near the surface of the slide.

Units operating in avalanche areas must carry, as a minimum, enough avalanche probes to equip one infantry squad. This will insure that immediate rescue operations can be undertaken should an avalanche occur and bury a victim.

Action of Individuals Caught in an Avalanche. When caught in an avalanche, there are certain actions that can be taken which will greatly increase chances of survival and recovery. Most important, *DON'T PANIC.* Maintain self-control and attempt to stay on the surface and get out of the main slide path. Execute *"swimming"* motions to combat the tendency of the sliding, rolling snow to pull one underneath. *Grab any obstruction* to assist the fight to the surface. *Remove skis* if possible. *After being completely covered by the snow mass, both arms are held in front of the face and pushed against the covering snow in an attempt to provide an airspace before the slide hardens.*

Selection of Tactical Locations. Great care must be exercised in the selection of battle positions, command posts, outposts, security elements, battery positions, bivouac sites, and administrative support bases in the areas where hazardous conditions exist.

3—12. Avalanche Rescue

The following rescue procedures are based on approved and tested techniques developed by the Alta Avalanche Study Center, US Department if Agriculture Forest Service. Military rescue operations, therefore, differ only slightly from civilian operations. The main difference being the immediately available surviving personnel left at the avalanche site. These can be portions of a squad, platoon, or company. In civilian avalanche rescues, surviving personnel will generally consist of only one or two people. Because of this the odds of recovering a live victim in military operations are considerably higher than those in a civilian rescue attempt.

Survival Margin. If an individual is caught by an avalanche, prompt and organized rescue operations offer the only hope of getting the victim out alive. There are records of persons who lived as long as 72 hours while buried. Ordinarily, the victims are either killed instantly by crushing, or die within a short period from exposure, shock, and

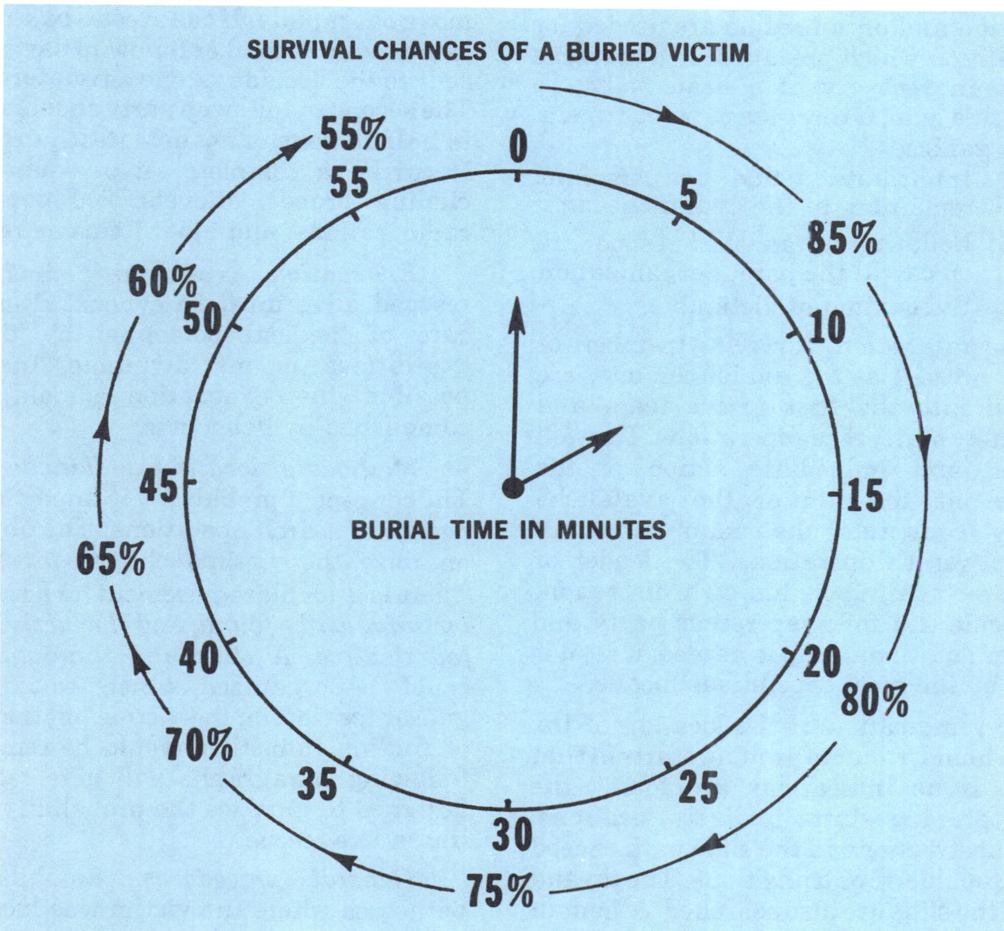

Figure 3—11 *Survival chances of a buried victim.*

suffocation. *Suffocation is the most important item.* Based upon experiences in the past, it can be concluded that *one hour is the average survival limit.* Snow is porous and ordinarily contains enough air to support life, though not consciousness. Statistics show that approximately 50% of the victims buried in an avalanche suffocate in an hour or less (fig. 3—11). Rescue operations are therefore designed to get the victim out within one hour. Because of some special circumstances which may prolong the life of the victim—he may be in an air pocket—rescue operations should not be discontinued for 24 hours at least.

Phases of rescue operations. The object of a rescue action is speed. Equipment and manpower must be assembled as quickly as possible at the avalanche site. Regardless of the rescue plan to be followed, exacting preparation and organization are needed for any operation which presupposes live rescue rather than recovery of a dead body. To achieve this goal, a three-stage rescue operation is organized:

- 1. Immediate action by personnel remaining at the avalanche site.
- 2. Followup by an alert rescue unit on call at the parent organization.
- 3. Evacuation of victim(s).

Immediate action. Surviving members of the unit, present at the avalanche site, are organized into the first rescue team and immediately start rescue operations. The well organized and immediate action of the personnel at the site of the avalanche generally constitute the main factor in successful rescue operations. The leader of this rescue team, upon his own discretion, must decide if a followup rescue party and any other additional help is needed; if so, it is sent for by the most expeditious means.

If any indication of the location of the victim is found, random probing starts in that vicinity. If no indications are found, the random probing starts in likely locations, such as obstructions in the slide path caused by trees, boulders, or transitions. The tip and edges of the slide are also searched. A human body is bulky and all being equal, is apt to be thrown toward the surface or the sides.

If the victim is found and is unconscious, first aid (artificial respiration) must be administered at once.

The alert rescue party, upon arrival at the scene reinforces the first rescue team.

Followup. The followup rescue party on duty, upon being alerted, is issued special equipment and provided transportation. This party is dispatched under command of the most experienced officer or noncommissioned officer available. The strength of the party should be at least one squad or, preferably, one platoon, and should include medical personnel. This party should be on their way within 15 minutes after receiving the alarm. This rescue party moves out lightly equipped; speed is the first consideration.

The strength of the alert rescue party is usually small to enable them to get ready and move out rapidly. If the area to be searched is extensive, a second or followup party must be sent to the accident scene as reinforcements. The second or followup party should move out in half an hour, or not not later than one hour. It carries a complete set of equipment including probes, shovels, 200 pound sled, radio, rations, and other items as required.

Evacuation. Avalanche victims who are rescued alive must be evacuated under the care of medical personnel by the most expeditious means available. Insofar as possible the evacuation should be accomplished by helicopter.

Methods of locating avalanche victims. The concept of probability is important to the design of search operations. The object is to optimize the victims chance for survival. There is a double requirement for success. *The victim must be found,* and *the victim must be found alive.* A slow and thorough search could be organized which would almost guarantee finding the victim, but the chances of finding him alive would be remote. The following paragraphs will give techniques designed to increase the probability of effecting a live rescue.

General procedures. Establish from witnesses where the victim was located just prior to the avalanche release, then determine the point where the victim disappeared—the "last seen" point. Making use of this and any other information, establish a probable victim trajectory line leading to high priority search areas. Make a rapid but systematic check of the avalanche debris surface and mark all clues. If justified by visible signs of equipment or avalanche cord, make an initial random probe of high-priority regions. If not justified, make a coarse probe of all likely areas of burial. Repeat the coarse probe as long as a live rescue remains possible. Resort to the fine probe only when the possibility of a live rescue is highly improbable.

Establishing the victim's most probable location. In many respects a moving avalanche resembles a fluid. A human body, with a higher density than the flowing snow, would be expected to sink deeper and deeper into the avalanche. Many factors obscure this statement. Turbulence, influence of terrain, and the victims own efforts to surface himself all interact to determine the final burial position. Study of a large number of case histories leads to the following conclusions.

The majority of buried victims are carried to the place of greatest deposition, usually the toe of the slide.

If two points of the victims trajectory can be established a high probability exists that the victim will be near the downhill flow line

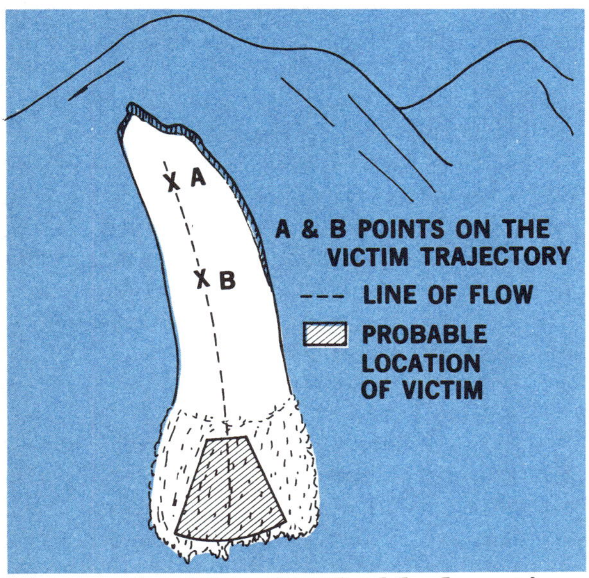

Figure 3—12 Probable location established by trajectory.

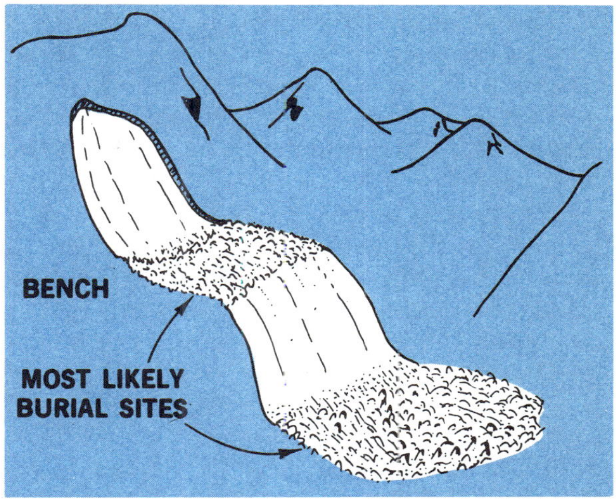

Figure 3—13 Effect of terrain on probable location.

passing through these two points (fig. 3—12).

Any terrain features which catch and hold avalanche debris are also apt to catch a victim (fig. 3—13).

If an avalanche follows a wandering gulley, all bends that show debris are likely burial spots. The likelihood of a victim being buried in a particular bend is proportional to the amount of debris deposited there (fig. 3—14).

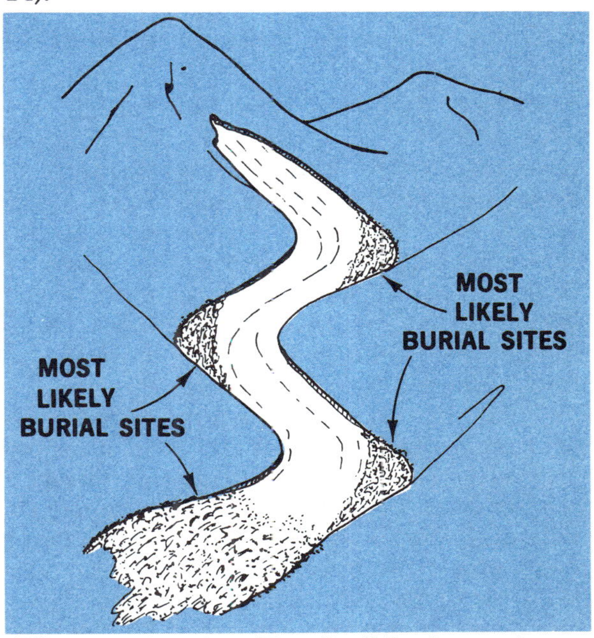

Figure 3—14 Probable locations in a wandering gulley.

398-868 O - 83 - 6

Vegetation, rocks and other obstacles act as snares. The victim tends to be retained above the obstacle. An obstacle may simply delay the victim's motion, leading to final burial down stream from the obstacle.

Maximum speed of the flowing snow occurs at the avalanche center. Friction reduces flow velocity along the edges. The closer the victim's trajectory is to the center of the slide, the greater will be his burial depth.

Efforts of the victim to extricate himself by vigorous motion and "swimming" definitely minimize burial depth. Conversely the limp body of an unconscious victim is likely to be buried deep.

An occasional exception to the above is emphasized. The victim may not be buried but may have been hurled away from the avalanche by wind blast. In the case of large and violent avalanches, a search of the surrounding terrain is advisable.

Probing for avalanche victims. Probing offers the advantage of requiring very simple equipment that can be operated by personnel without previous training. Although the probers do not need previous training the search leader must be familiar with the technique to insure proper execution of the probe line.

Probe Poles. Rigid steel tubing approximately 1.25 cm in diameter is recommended for the primary probe pole. Lengths of 3 to 4 meters are most convenient. Longer poles are difficult to manage especially in a high wind. Standardizing the length of poles in a given unit should be mandatory. When the poles are of equal length, it is immediately obvious when one probe in the line is stopped by an obstruction.

Although the rigid steel probe poles are the preferred, sectional poles should be carried by all units habitually operating in avalanche areas. This provides an immediate capability to begin probing by surviving members of the unit. In the absence of any kind of probe pole, initial probing attempts can be started using ski poles. This can be accomplished in two ways, the ski pole can be reversed, probing with the wrist strap down, or the basket can be removed, the preferred method, so that the point is down. This method allows the ski pole to penetrate the snow more easily than in the first method.

Precautions should be taken to prevent the probes from icing when large temperature variations are present. Probes should be kept moving in and out of the snow. There is some advantage in waxing the poles with ski wax. Probes should not be left fully inserted in snow because they will invariably freeze in position and will be exceedingly difficult to extract.

For the probing operation to be effective, lines must be orderly and properly spaced. To insure systematic and orderly probing the number of personnel per line should be limited. Twenty per line is satisfactory while thirty is the upper limit. The number of men in the line will be dictated by the width of the area to be probed. A stringline must be used to keep the probe lines aligned.

The probe line maintains a steady advance upslope. Advancing uphill automatically helps set the pace and permits easy probing to the full length of the probe. Probing does not come to a halt when a possible contact is made. The probe is left in contact and the line continues. A shovel crew follows up on the strike by digging down along the pole. Extra probes are carried by the shovel crew to replace those left in contact. Such a plan of operation is especially important when more than one victim is buried.

Striking a body gives a distinct feel to the probe. This feel is easily recognizable in soft snow but is less so in hard compacted snow. A common problem is encountering debris within the snow that can be mistaken for the victim. The only sure check is by digging.

Probing Techniques. Two distinct probing methods are recognized :

- Coarse probe
- Fine probe

As evidenced by their names, coarse probing implies a wider spacing of probe pole insertions with emphasis on speed. Fine probing involves close-spaced probing with emphasis on thoroughness. Coarse probing is used during initial phases of the search when live recovery is anticipated. Fine probing is the concluding measure which almost guarantees finding the victim. Effectiveness of these two probing techniques can be compared by examining the relation of sampling (probe insertion) intervals to the projected area of the human body. The result is that the coarse probe technique has a 76% chance of locating a victim on a given pass, while the fine probe has essentially a 100% chance of locating the victim.

The Coarse Probe. The coarse probe functions as follows:

- Probers are spaced along a line 75 cm center to center. A distance of 50 cm is straddled, leaving 25 cm between the toes of adjacent probers (fig. 3—15).
- A single probe pole insertion is made at the center of the straddled span.
- On signal of the probe line commander, the group advances 60 cm and repeats the single probe.
- Two signals are used for the complete sequence "DOWN PROBE" and "UP PROBE." By adhering to these commands the leader can look along the probe line when the probes are down and instantly spot a "high" probe which might indicate a possible location for the victim. It is important that the signals be adjusted to a rhythm which enforces the maximum reasonable place. Strict discipline and firm, clear commands are essential for efficient probing. The probers themselves work silently.

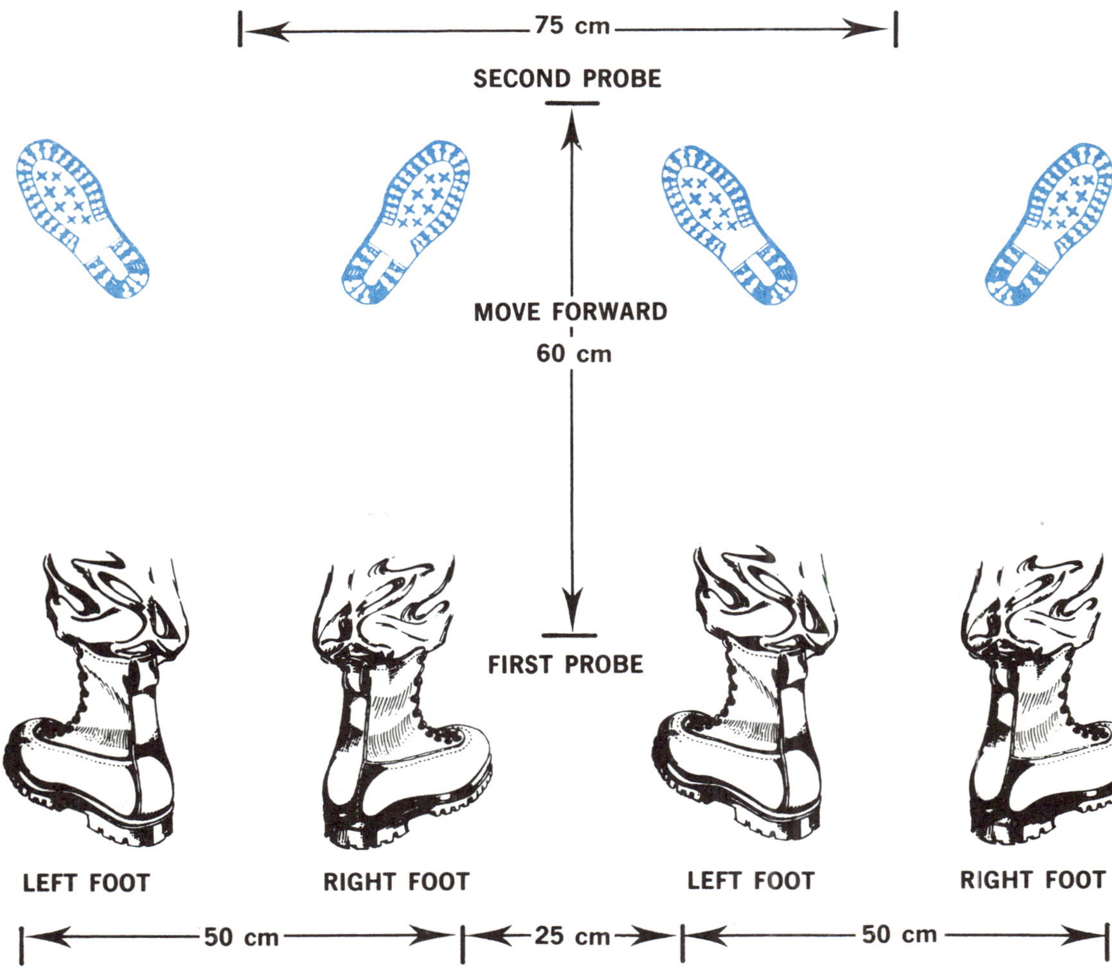

Figure 3—15 ***Coarse probe.***

The Fine Probe. The fine probe functions as follows :

- Probers are spaced the same as for the coarse probe.
- Each man probes in front of his left foot, then in the center of his straddled position, and finally in front of his right foot.
- On signal, the line advances 30 cm and repeats the probing sequence (fig. 3—16).
- The exact number of signals from the probe line commander is arbitrary, but the fine probe functions best when each probe pole insertion and extraction is done on command.
- Good discipline and coordinated probing is even more necessary than with the coarse probe. Careless or irregular probing can negate the advantages of fine probing. Use of a stringline to aline the probers is especially important with the fine probe. The three insertions are made along the line established by the stringline which is then moved ahead 30 cm.

Shovel Crews and Trenching.

- Shovel crews are organized to assist the probers, relieving them at intervals and digging in likely spots. They should be prevented, however, from haphazard digging and wasting energy.
- Time spent in fine probing should not exceed 3 to 4 hours. If this is unsuccessful, the slide must be trenched by the reinforcing elements.
- Trenches are dug parallel to the contour down to ground level or to undisturbed snow at intervals of 2 meters. Digging begins at the tip of the slide and proceeds uphill. It is feasible to space the shovel teams along one

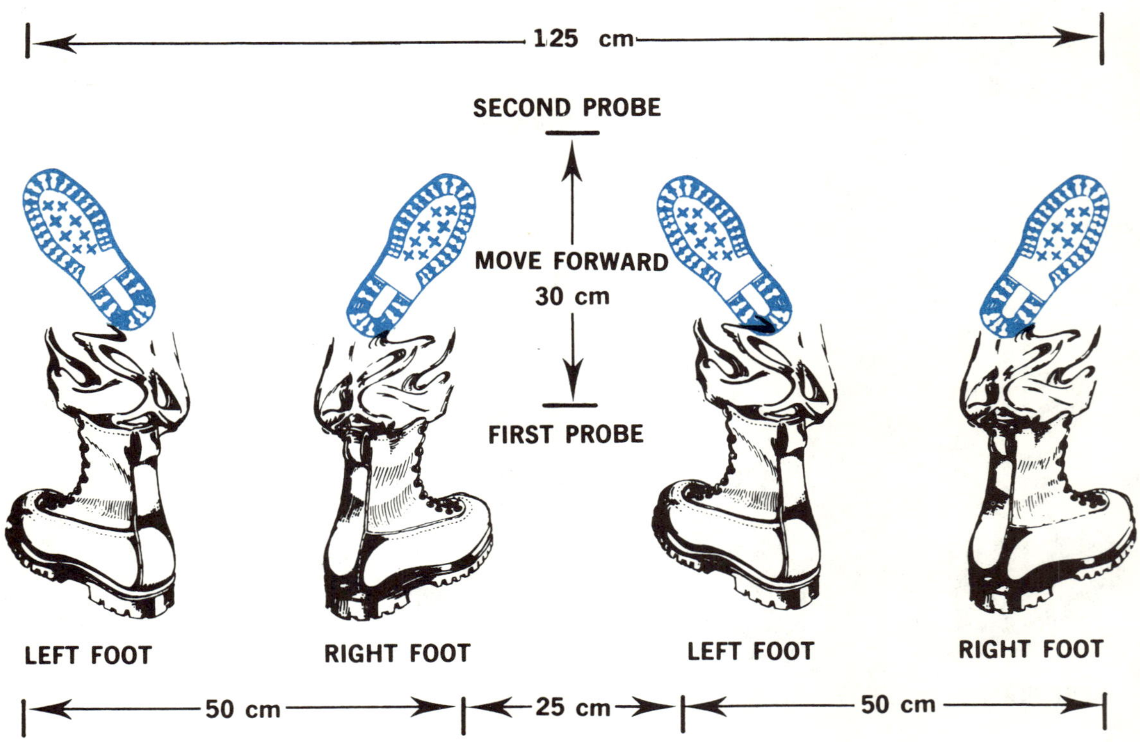

Figure 3—16 ***Fine probe.***

trench with frequent intervals. In this manner snow from one trench can be thrown into the one just completed.

- If trenching is necessary, the operation ceases to be of an emergency type. Trenching should be resorted to only if deemed in the unit/command's best interests. Trenching will recover the remains; however, the dangers inherent at the avalanche site plus the exceptional effort that trenching requires should be considered in light of the ability to recover the remains after the spring or summer thaw.

Avalanche Rescue Equipment. The following rescue equipment is considered the minimum for organizations of battalion size; it should be readily available to the alert force at all times.

20 probes (sectional preferred—otherwise metal or wooden rods 1.25 cm in diameter, 3 to 4 m long).

12 snow shovels, or D-handle GP shovels.

12 flashlights with spare batteries.

6 nylon climbing ropes, 36½ meters long.

6 colored avalanche cords, 20—30 meters long.

2—200 lb sleds, with harnesses and tow ropes.

2 emergency evacuation bags or 4 sleeping bags, arctic.

1 set of first aid equipment.

Emergency rations.

Stoves and cooking equipment.

Skis, ski poles, and climbers for each individual.

Tentage.

Platoon and company radios.

20 marking flags (brightly colored).

Section IV MOVEMENT OVER SNOW AND ICE

3—13. General

Movement over snow with the aid of skis and snowshoes is discussed in detail in FM 31—70.

The principles which apply to movement over rocky terrain also apply to over-snow travel when skis or snowshoes are not used. When snow lies over a foundation of other than glacier ice, and the terrain is not steep, no special techniques are called for. Efficient progress depends largely upon proper route selection and on using procedures that will minimize exertion.

3—14. Equipment

Ice ax (fig. 3—17). The ice ax to the trained mountaineer is one of the most important pieces of his equipment. Primary uses are discussed in the proceeding paragraphs; however, each man will find, through experience, other uses for the ax. Extreme care must be exercised at all times in handling and using the ax. If carelessly handled the ice ax becomes a dangerous weapon, not only to the owner but to others as well.

Primary uses:

- Cutting steps in ice and snow.
- Probing for snow covered crevasses and obstructions.
- Belaying in ice and deep snow.
- Aid in steep terrain (third point of suspension).
- Implement to stop falls (self arrest).

Carrying. On fairly level open trails the ice ax may be carried by grasping it at the balance with the head to the rear, pick down.

It may also be used as a walking stick. When walking on snow, the head is grasped with the pick pointing forward; when walking on ice the adze is pointed forward. Alternately, the ice ax may be carried under the arm, in this method the head is to the rear, parallel to the ground with the pick pointing away from the body. If no need exists for the ax, it may be stowed in the rucksack head down with the shaft protruding.

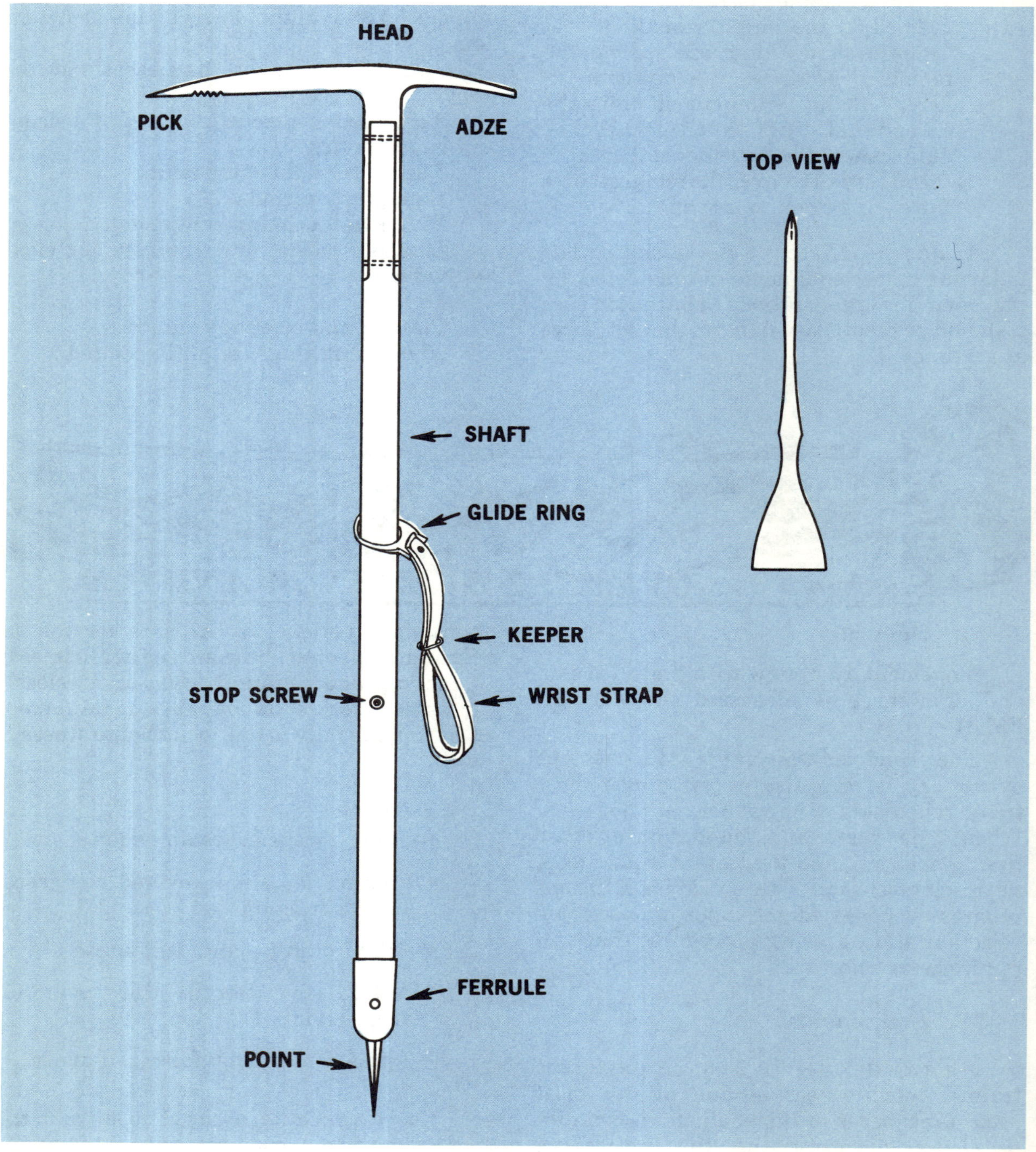

Figure 3—17 *Ice ax.*

Crampons (fig. 3—18). The crampon is used as an aid in securing additional traction on snow and ice. Care must be exercised to insure proper fit. The crampon should fit snugly to the boot. The front points should neither protrude past, nor fail to reach the toe of the boot.

Technique. Balance is important, the ankle should be flexed so that the crampon points are at right angles to the slope, this applies in both the ascent and descent. On hard snow the points of the crampons will bite in easily, but on hard ice the foot should be stamped down hard to drive the points in.

Maintenance. The crampons should be inspected frequently to make sure straps and rivets are secure and serviceable. The points should be kept filed sharp.

Field expedients. The crampon may tend to break at the hinge where the front and rear spikes join. As an expedient, they can be wired together and used until replacements are obtained. In an emergency the two sections can be joined together with a snaplink.

Safety. Care should be exercised when wearing crampons to insure that they do not

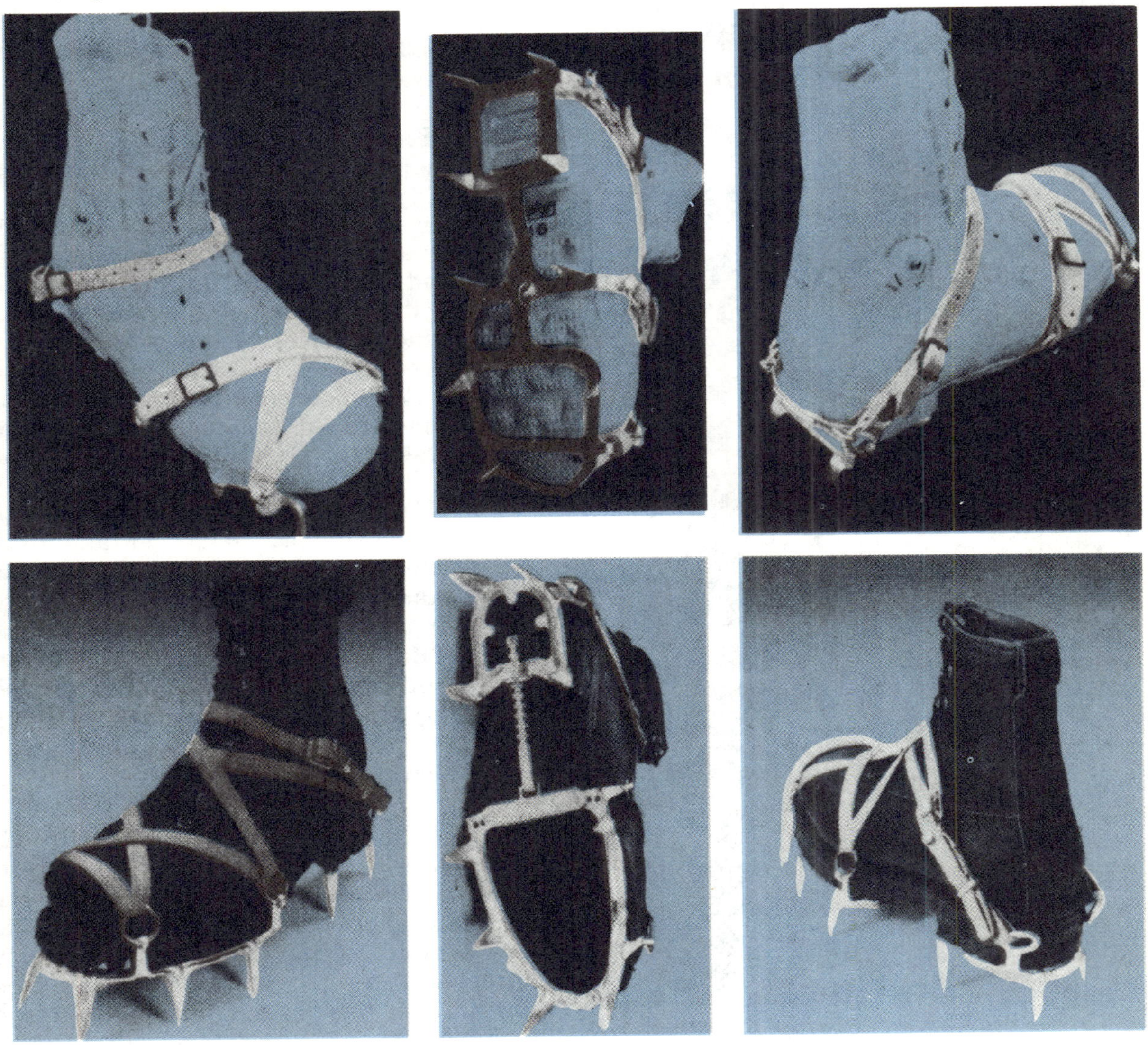

Figure 3—18 ***Crampons.***

catch in the trouser leg when climbing.

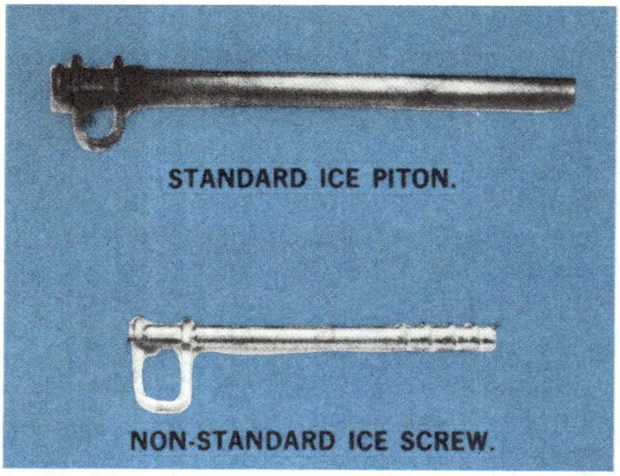

Figure 3—19 Ice piton and ice screw.

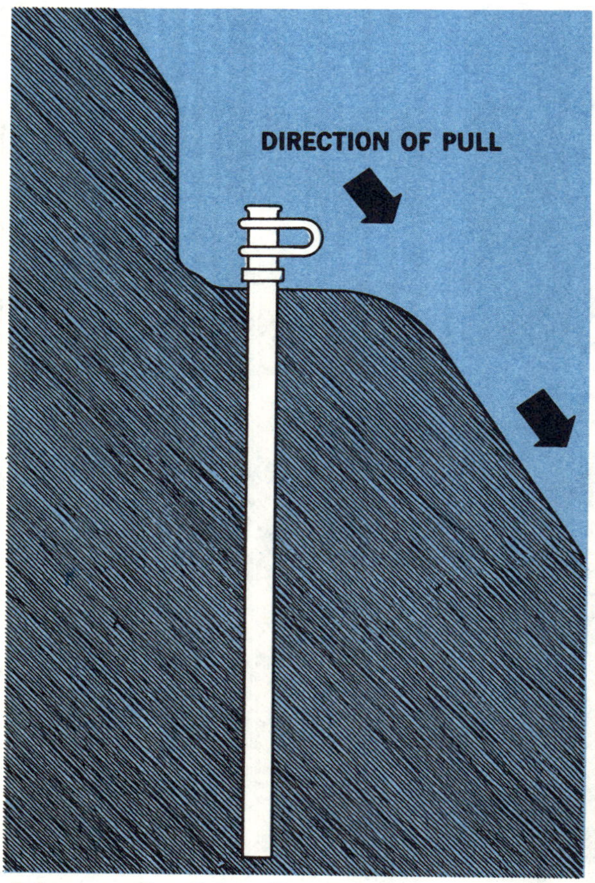

Figure 3—20 Proper method of emplacing the ice piton.

Ice Anchors.

Ice pitons and ice screws (fig. 3—19). Ice pitons and ice screws are used to establish anchor points. The ice piton is a standard item of issue while the ice screw is a nonstandard commercial item. Emplacement and use of the two anchors is identical with the ice piton being driven into the ice and the ice screw turned. The ice piton is emplaced as follows:

- Cut a horizontal recess in the ice and remove all rotten ice.
- Drive the piton vertically all the way to the ring. Applied or expected pull should be at right angles to the piton shaft (fig. 3—20).
- Test the piton before putting pull on it to insure that it is properly emplaced (fig. 3—21). If the piton pulls out or appears weak, move to another spot and redrive it. In most cases it will be advisable, for safety, to employ the pitons in pairs. In this case the pitions should be driven at right angles to each other (fig. 3—22). Employing the pitons in this manner prevents either one of them from pulling out accidentally.
- Use time of the ice piton is limited, the piton will heat from solar radiation or the ice will crack or soften. The use time can be extended by covering the pitons or screws with ice chips once they have been emplaced. If considerable climbing is to be made utilizing ice pitons, they should be inspected frequently and redriven when necessary.

Horseshoe anchor (fig. 3—23). The horseshoe anchor is a horseshoe shaped projection fashioned from ice or snow and constructed by either cutting with the ice ax or stamping with the boots. When constructed of snow, the width should be not less than 1.5 meters. In ice, this width may be narrowed to .5 to 1 meter, depending on the strength of the ice. The trench around the

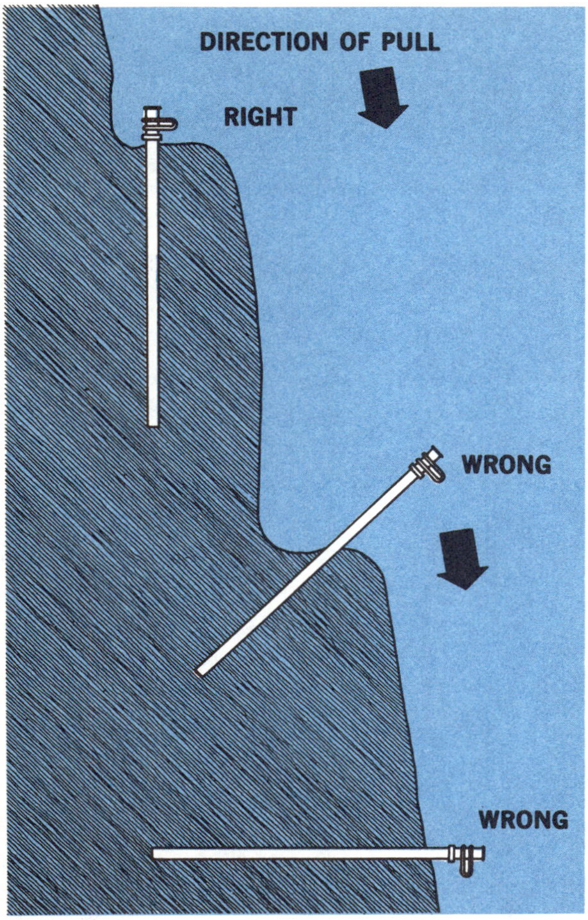

Figure 3—21 ***Right and wrong way of emplacing the ice piton.***

horseshoe should be stamped as deeply as possible in the snow and should be cut to not less than 15 cm in ice. The back side of the anchor must *always* be undercut to prevent the rope sliding off and over the anchor. A snaplink is attached to the front of the anchor to prevent the ropes from spreading.

The properly constructed horseshoe anchor provides an anchor point always available and may be used for rappels and for fixed ropes.

The horseshoe should be inspected frequently to see that the rope is not cutting through the snow or ice. If such is the case, a new anchor should be constructed in a new location.

A horseshoe anchor constructed in snow is always precarious and its strength depends upon prevailing texture of the snow. For dry snow or windpacked snow the reliability of the anchor is questionable.

3—15. Movement Over Snow and Ice

On very steep slopes in deep snow, the climber may climb straight up facing the slope. The ice ax driven directly into the snow provides a quick and effective protection in case of a slip. It is usually best, however, to climb snow slopes in a zigzag fashion in order to conserve strength. If the snow is packed or frozen hard, the sole of the boot will generally hold by kicking steps, even in steep slopes. Where it is difficult to make an effective step with the boot, a cut with the adze end of the ice ax makes an effective step. In these situations crampons should be used for faster and easier movement.

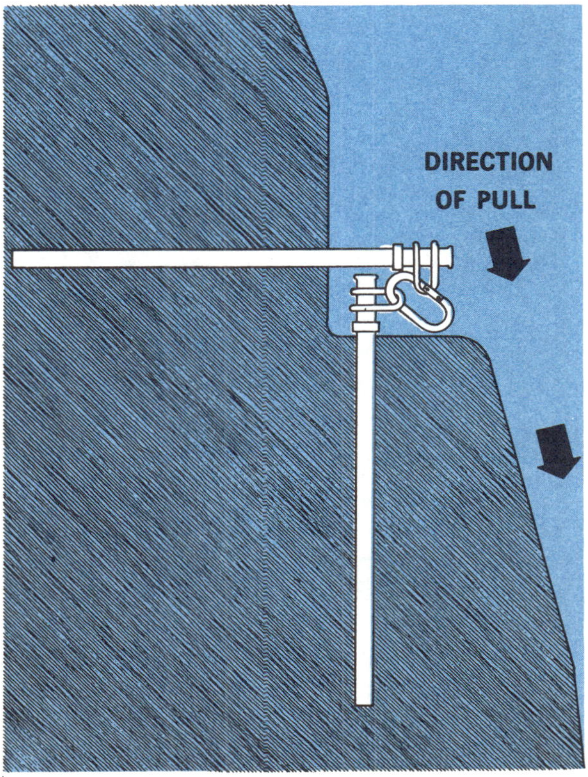

Figure 3—22 ***Ice pitons emplaced at right angles.***

SIDE VIEW

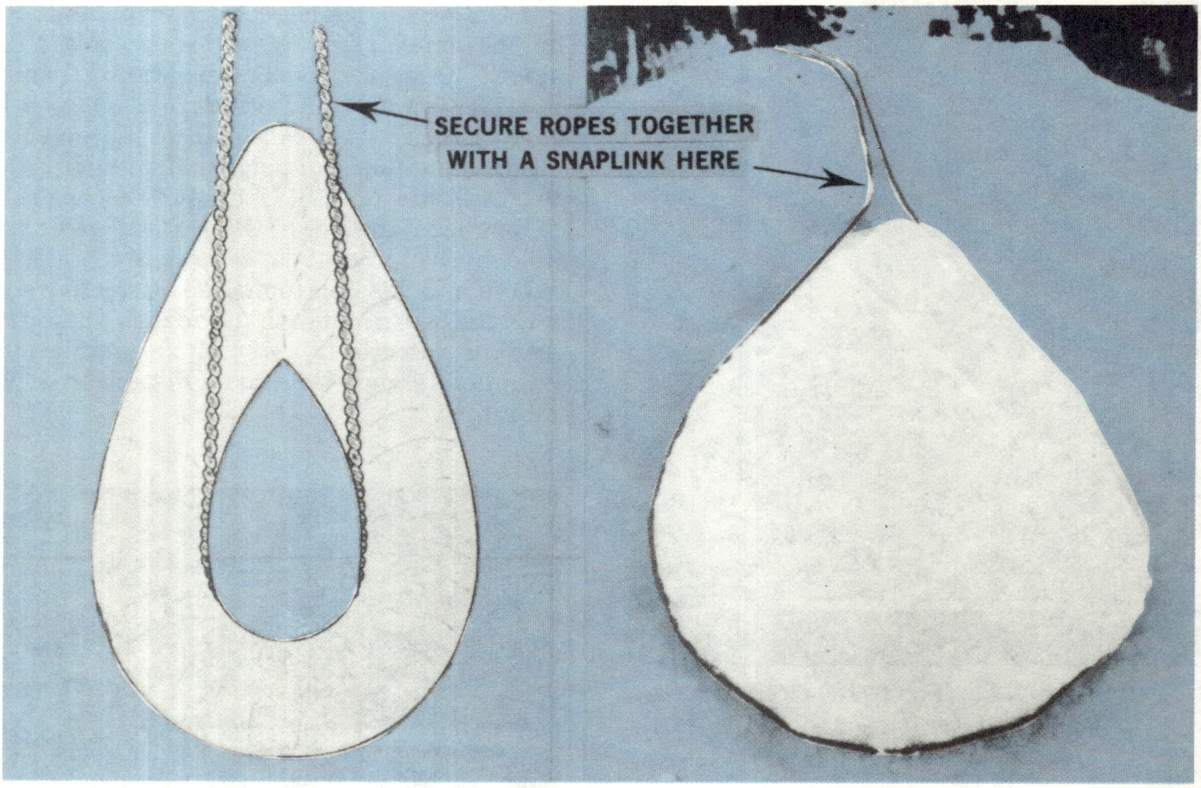

TOP VIEW

Figure 3—23 ***Horseshoe anchor.***

When descending in snow, one can usually come straight down even in steep terrain. Movement downhill should be slow and deliberate and with an even pace. The heels should be kicked vigorously into the snow wherever it is hard packed. The body may be kept erect with the help of an ice ax, which may be jammed into the snow at each step for additional safety. Here again, crampons or step-cutting may be necessary.

Step cutting in ice (fig. 3—24) is done with with pick end of the ice ax. A blow directly in and another diagonally toward the first will usually fashion a rough step which may be finished off with the adze end of the ice ax. Black (dark) ice is harder and requires more blows. The blows should be directed into the ice at as near a right angle as possible to reduce the danger of flaking off the outside layers. The step should slope slightly inward. It does not need to be large except at resting places. In climbing, steps may be cut straight up or on a diagonal to the line of ascent. In descending it is seldom practicable to do anything except to cut steps in a diagonal line.

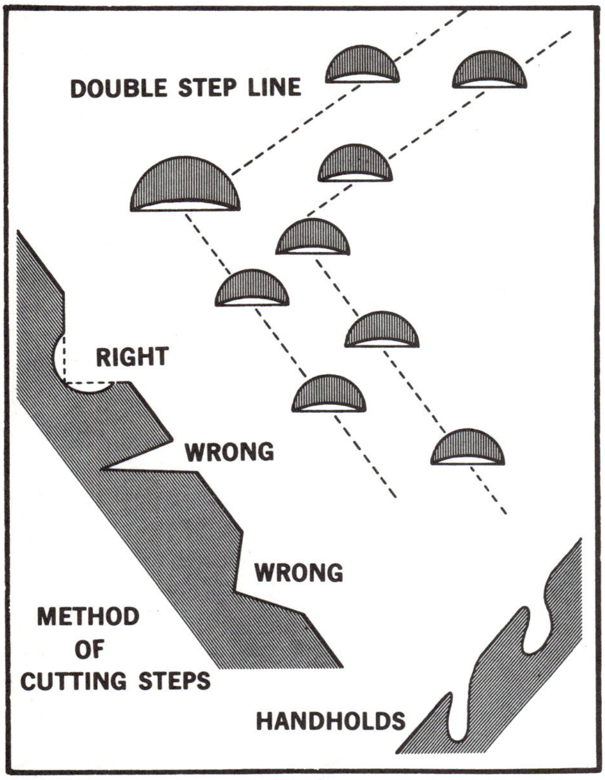

Figure 3—24 Step cutting in ice.

Belays are more difficult to secure on ice and snow than on rock, but they are equally necessary. In general, they are obtained by use of the ice ax although ice pitons are also used; for rappelling, a belay point can be cut directly into the ice itself.

On hard packed snow slopes which can be seen from top to bottom and which are not too steep, it is possible to *glissade* (fig. 3—25). *Glissading should not be attempted unless the bottom of the slope can be seen,* otherwise unseen obstacles may result in serious falls. Glissading in a sitting position gives little control of descent. In the standing position, the body must always be kept erect, as nearly upright as possible, and with the ice ax firmly held in the hands. The ice ax is pressed back into the snow when necessary. Turning the toes to one side or the other changes the course of descent. Turning sharply to the side on which the ice ax is held increases the pressure on the ice ax and brings one to a stop. To slow down, pressure of the ice ax is sufficient. *Glissading down solidly frozen ice slopes is dangerous, and should not be attempted.*

Figure 3—25 Glissading.

Section V MOVEMENT OVER GLACIERS

3—16. Principal Dangers

The principal dangers and obstacles to movement in glacial areas are crevasses, icefalls, and ice avalanches.

Snow covered crevasses make movement on a glacier very treacherous. In winter, when visibility is poor, the difficulty of recognizing them is increased. Toward the end of the summer, crevasses are widest and covered by the least snow. Snow bridges constitute the greatest potential danger in movement over glaciers.

On the steep pitch of a glacier, ice flowing over irregularities and cliffs in the underlying valley floor cause the ice to break up into ice blocks and towers, criss-crossed with crevasses. This jumbled cliff of ice is known as an *icefall.* Icefalls present a major obstacle to safe movement of troops on glaciers. Ice avalanches are common in the vicinity of icefalls and can threaten men on or at its base. The safest time for movement through an icefall is early in the morning before sunrise.

Light and heat rays reflected from ice, snow, water, and bright rocks irritate and burn the skin very rapidly, causing glacial sunburn. Sunburn can occur even on cloudy days. A strong wind will make the burn more severe. Sunburn cream should be used frequently on all exposed skin surfaces. It is particularly dangerous to expose parts of the body which are not accustomed to the sun's rays. As soon as any part of the body becomes burned, it should be protected from further exposure. Particularly bad cases of burn very often lead to fever and also to a lessening of muscular activity. It may often take several days before the casualty is fit for duty.

Snow blindness occurs when sun is shining brightly on an expanse of snow, and is due to the reflection of ultraviolet rays. It is likely to occur after a fall of new snow, even when the rays of the sun are partially obscured by a light mist or fog. In most cases, snow blindness is due to negligence or failure on the part of the soldier to use his goggles. Symptoms of snow blindness is a sensation of grit in the eyes, pain in and over the eyes, watering, redness, headache and distaste for light. First aid includes blindfolding the eyes for a few days until the pain is gone. Although there is no permanent damage, personnel will be out of action until the eyes recover.

Unforeseen falls are always a possibility when moving over ice or snow covered slopes. If the party is roped together, the faller can usually be arrested by other members; however, if the party is traveling unroped, the *self-arrest* system using the ice ax must be employed. After falling, roll immediately onto the stomach in direction of the head of the ice ax. If wearing crampons, the legs spread apart and bent at the knee with the

Figure 3—26 Self-Arrest.

feet up in the air; if crampons are not worn, the feet may be dug in to assist in the arrest. The head of the ice ax is held with one hand shoulder high; the shaft of the ax is grasped with the other hand and held diagonal to the body. The ax is secured to the hand with the wrist strap. To arrest the fall, the head of the ice ax is pressed into the surface with increasing pressure until movement is arrested. If the fall occurs on ice, the pick end of the ice ax should be used on the surface; if on snow the adze end should be used (fig. 3—26).

3—17. Use of Rope

A roped party of two, while ideal for rock climbing, is at a great disadvantage on a snow-covered glacier, since it is almost impossible for one man alone to rescue another who has fallen into a crevasse. *The best combination is a three-man party, tied in*

Figure 3—27 Method of carrying rope coil when moving on glaciers.

to a single rope. Each member should carry, in a coil, enough of the rope to eliminate a direct pull on his body loop. The rope should touch the snow as little as possible, since a wet rope kinks badly and may freeze and become unmanageable (fig. 3—27). It is best to travel roped at all times when on the snow-covered part of a glacier. Use of the rope on steep and/or heavily crevassed ice, however free of snow, is *mandatory.*

3—18. Routes

The individual operating in the mountains must appreciate certain limitations imposed by nature in glacier movement. Access to the end portion of a glacier may be difficult due to abruptness of the ice and possible presence of crevasses. Additional obstacles in getting onto a glacier may be swift glacier streams, steep terminal or lateral moraines, and abrupt mountain terrain bordering the glacier ice. The same obstacles may also have to be overcome in getting on and off a valley glacier at any place along its course. Further considerations to movement on a glacier are steep sections, heavily crevassed portions, and icefalls, which may be major obstacles to progress. The use of up-to-date aerial photographs in conjunction with aerial reconnaissance is a valuable means of gathering advance information about a particular glacier; however, they only supplement and do not take the place of on-the-ground reconnaissance conducted from available vantage points.

3—19. Crossing Crevasses

Open crevasses are obvious, and their presence is an inconvenience rather than a danger to movement. Narrow cracks can be jumped, provided the takeoff and landing spots are firm and offer good footing. Wider cracks will have to be circumvented unless a solid piece of ice joins into an ice bridge strong enough to support at least the weight of one member of the party. Such ice bridges are often formed in the lower portion of a crevasse, connecting both sides of it.

In the zone that divides seasonal melting from permanent falls of snow, large crevasses remain open, though their depths may be clogged with masses of snow. Narrow cracks may also be covered. In this zone, the snow which covers glacier ice melts more rapidly than that which covers crevasses. The difference between glacier ice and narrow snowcovered cracks is immediately apparent; the covering snow is white, whereas the glacier ice is gray.

Usually, the upper part of a glacier is permanently snow covered. The snow surface here will vary in consistency from dry powder to consolidated snow. Below this surface cover are found other snow layers that become more crystalline in texture with depth, and gradually turn into glacier ice. It is in this snowcovered upper part of a glacier that crevasses are most difficult to detect, for even wide crevasses may be completely concealed by snow bridges.

Snow bridges are formed by windblown snow which builds a cornice over the empty interior of the crevasse. As the cornice grows from the windward side, a counter part is formed on the leeward side. The growth of the leeward portion will be slower than that to the windward so that the juncture of the cornices occurs over the middle of the crevasse only when the contributing winds blow equally from each side. Bridges can be formed also without wind, especially during heavy falls of dry snow, and since cohesion of dry snow depends only on an interlocking of the branches of delicate crystals, such bridges are particularly dangerous during the winter. When warmer weather prevails during the summer, the snow becomes settled, is more compacted, has greater carrying capability, and forms firmer bridges.

Once a crevasse has been completely bridged, its detection is difficult. Bridges are generally slightly concave because of the settling of the snow. This concavity is perceptible in sunshine, but impossible to detect on dull days or during a snowstorm. If the presence of hidden crevasses is suspected, the leader of a roped party must probe the snow in front of him with the shaft of his ice ax (fig. 3—28). As long as a firm foundation is encountered, the party may proceed, but should the shaft meet no opposition from an underlying layer of snow, a crevasse is

Figure 3—28 Probing for a crevasse.

probably present. In such a situation, the prober should probe closer to his standing place to make sure that he is not standing on the bridge itself. If so, he should retreat gently from the bridge and determine the width and direction of the crevasse. He should then follow and probe the margin until a more resistant portion of the bridge is reached. When moving parallel to a crevasse, all members of the party should keep well back from the edge and follow a course parallel to it.

A crevasse should be crossed at right angles to its trend unless a diagonal bridge is clearly visible. When a bridge which seems sufficiently strong to hold a member of the party has been located, proceed as follows:

- The leader and second man takes up position at least 3 meters back from the edge. The third man goes into a belay immediately behind the second man.
- The second man belays the leader by plunging the shaft of his ice ax firmly into the snow surface, taking a loop around it with the rope, and paying out line as the leader advances. If the snow is too powdery to provide good anchorage for the ice ax, or belaying area is solid ice, the second man should sit in the snow with his feet braced in front of him and use the hip belay prescribed for rock climbing (fig. 3—29). The ice ax belay should only be

Figure 3—29 Method of establishing a belay for crossing a crevasse.

used in deep snow, and should not be attempted when the surface is hard ice.

- The leader should move forward, carefully probing the snow and evaluating the strength of the bridge, until he reaches firm snow on the far side of the crevasse.
- Having crossed the crevasse, the ladder should belay the second man in the same manner as described above.
- When the second man has crossed safely, the leader should move forward to a secure position while belays are provided for the remaining members of the party by each succeeding climber.

In crossing crevasses, distribute the weight over as wide an area as possible. Do not stamp the snow. Many fragile bridges have been crossed by lying down and crawling to the other side.

3—20. Crevasse Rescue

To provide a quick means of holding an unexpected breakthrough, the rope is always kept taut and is secured to the ice ax shaft with a round turn. When an unexpected breakthrough by the first man takes place, the second and third men thrust their ice axes immediately into the snow to arrest the fall. A fall through a snow bridge results either in the man becoming jammed in the hole broken by his feet, or in being suspended with the crevasse by the rope.

If the leader has fallen only partially through the snow bridge, he is supported by the snow forming the bridge. *He should not thrash about.* This will only enlarge the hole and result in deeper suspension. All his movements should be slow and aimed at rolling out of the hole and distributing his weight over the remainder of the bridge. Pulling on the rope with one or both hands is of great assistance. The victim can help himself more than he can be helped by the second man, whose duties consist primarily of belaying the rope firmly to check any further breakthrough on the leader's part. It generally is safer to retain the rucksack, as its bulk often prevents a deeper fall. Should a man other than the leader experience a partial fall, the rescue procedure will be the same as for the leader.

When the man is suspended by the rope, the length of his fall depends upon how quickly the rope can be checked and where in the bridge the break takes place. If the fall occurs close to the near edge of the crevasse, it usually can be checked before the climber has fallen more than 1 or 2 meters. However, if he is nearly across, the fall will cause the rope to cut through the bridge, and then even an instantaneous check by the second man will not prevent a deeper fall.

In such a fall, it is necessary to act

promptly to relieve the constriction of the loop around the victim. The victim can ease the strain by lifting himself on the rope, but he can do so for only a very short period. A stirrup constructed from a sling rope and secured to the belay or climbing rope by a prusik knot can be used as a rest loop to east the constriction of the rope around the waist. This stirrup should be made up ahead of time, attached to the climbing rope, and tucked into a pocket until needed. This method also frees the hands, and enables the man to assist in further rescue operations. Personnel on the surface must ascertain the conditions existing in the crevasse. For example, when firm masses of snow clog the crack some distance below the surface, it is easier to lower the victim to such a platform and begin rescue operations from there. In estimating the conditions within the crevasse, it should be remembered that it is almost impossible to hear the voice of the victim even a short distance from the hole through which he fell. Therefore, one member of the party, while being belayed by another, must approach the hole in order to gain the required information.

Since a snow bridge is usually strongest at the edge of the crevasse, a fall is most likely to occur some distance away from the edge. In such places, the rope usually bites deeply into the snow, thus greatly increasing friction for those pulling from above. In order to reduce friction, place an ice ax or ski under the rope and at right angles to the stress. If a ski is used, place the running surface down so that

Figure 3—30 Ascending from crevasse by use of stirrups.

the rope will not cut on the edges. Push the ice ax or ski forward as far as possible toward the edge of the crevasse, thus relieving the strain on the rope. If the victim is able to reach firm footing within the crevasse, the support may be placed forward at the most advantageous point.

The rescue procedure remains essentially the same, whether or not the victim is able to reach firm footing within the crevasse. Spare rope should always be carried by a single rope party. If no spare ropes are available, as many members of the party as necessary will have to unrope until he can reach it; it is then securely belayed on the surface. While he is being pulled up by this rope, he assists by attempting to pull himself up on the anchored rope. His progress is aided by the rescuers taking up the body line. Whenever possible, the party or parties following should move up to assist in rescue operations.

An alternate method for use by a three-man team is for the number three man to unrope, tie a bowline loop on the end of the climbing rope, and lower it to the man in the crevasse. The fallen man, meanwhile takes out the prusik stirrup that he previously tied onto the climbing rope, and places it down through the bowline around his waist, between his legs, around the outside of the leg, and over the toe of the boot. By moving the prusik up the climbing rope about 50 cm, he can then stand in the stirrup, thus taking the strain off his waist and chest. The bowline loop, lowered from above, is placed on the other foot in a like manner. The fallen man then uses the stirrup and bowline loop as steps while being assisted by the belay man above. This provides an upward step for him. By a repetition of this procedure, alternating stirrup and bowline loop, the fallen man will approach the lip of the crevasse, (fig. 3—30). This method is standard for use by three-man teams if the fallen man is unable to reach either wall of the crevasse. The middle man must have two stirrups attached to the climbing rope in using this method. Stretch and elongation of the nylon climbing rope make this operation difficult and lengthy if the man has fallen more than 7 meters. If an overhang is encountered the best procedure is for him to grasp the fixed rope, place both feet against the snow or ice of the crevasse wall and "walk" out over the overhanging snow while the slack on the body line is taken in and belayed from above (fig. 3—31).

If the suspended climber is injured or unconscious and unable to help in his own rescue, additional assistance by one or more rope parties is then needed to recover the victim. To insure free movement and unnecessary strain on the lowered climber, a three-loop bowline knot is used. Two loops are made larger and one is passed over each leg, while the third and smaller loop is placed around his chest. To facilitate recovery operations and for additional safety, it is advisable to secure him with an additional

Figure 3—31 Ascending from crevasse by foot on a fixed rope.

398-868 O - 83 - 7

section of a climbing rope. As this climber is lowered, another rope which is also tied with a three-loop bowline is passed down to him. The victim is tied to this rope by the rescuer in the same way. The rescuer helps the victim as both men are pulled out of the crevasse by the belayers above. It will facilitate the recovery of the two climbers if the second rope securing the victim or the rescuer in the crevasse is managed from the opposite side of the crevasse where the belayers take up their positions in advance. In this manner victim and rescuer are pulled free from the crevasse wall. As they reach the top of the crevasse, the ropes on one side are slowly slacked off while both men are pulled over the opposite crevasse edge.

- When additional assistance cannot be secured, rescue operations are more difficult and actions taken depend mainly upon the circumstances. They must be made in the best interest of the injured man. It is most important that he is relieved from his suspended position as soon as possible.

  If the victim is injured so that he cannot use either leg in the stirrup method described above, the loop of the lower rope section is then made sufficiently large to be passed by the victim over his shoulders and around his chest. Secured in this manner, he is then pulled up by both belayers. When the victim is unable to move or is unconscious, possibilities of lowering him to a firm snow bridge or shelf immediately below him are investigated by the second belayer, while he is secured by the other climber. This possibility is only of advantage when additional help can be secured within a reasonable amount of time and the victim rescued. Prolonged exposure in such environment is prone to cause frostbite and shock in a relatively short period of time. If this procedure is not practicable, the victim is pulled cautiously out of the crevasse by both belayers. An attempt should first be made to lower a rope section with a large loop and pass it around the upper part of the victim's body in an effort to keep him in an upright position, and for additional safety.

Usually a leader who has been extricated from a suspended fall should not be permitted to immediately resume the responsibility of route selection. A free fall of only a few meters, and the resulting rescue operations, are a severe strain on a man.

Should it become necessary for a roped two-man party to traverse a glacier area, each man should have, already tied to climbing rope, three prusik slings made from sling ropes. One sling is worn around the chest and the other two are tied to form stirrups. If one man falls into a crevasse and is belayed by his companion, he places the stirrups down through the bowline around his waist, between his legs, around the outside of each leg, and over the toe of each boot. By weighting one foot, the fallen man relaxes the other stirrup, permitting the prusik knot to be pushed farther up the rope. Thus, by alternating this operation with each stirrup, and by pushing the chest loop knot even higher, the man extricates himself from the crevasse.

Chapter 4

FIXED ALPINE PATHS

Section I GENERAL

4—1. General

Fixed alpine paths are normally constructed when operations become stabilized and the commander can be assured that he will remain in one given area for a considerable period of time. They are also constructed when there is no other, or limited means, to move men, equipment, and supplies, to and from forward areas. Although these paths can be built by any troops in the area, they normally fall under the perview of assigned or attached engineer elements.

4—2. Purpose

A fixed alpine path is constructed of materials at hand to assist troops in traversing rugged mountain terrain, ascending or descending steep mountain slopes, and to aid in transporting supplies. It should be constructed in a manner suitable for use by men carrying combat equipment or supply loads up to 50 pounds. If possible, it should be constructed so that movement is possible in both directions at once and so that casualties can be evacuated over it on litters.

4—3. Description and Data

Description. A fixed alpine path is a combination of aids to movement, and may include steps cut into the ground or into the ice or snow, hand lines supported by stanchions (fig. 4—1) hanging ladders of wire or fiber rope up nearly vertical rock faces, stand-off type ladders made of local materials, suspended walkways to cross ravines, cable bridges, and hand-operated cableways (fig. 4—2).

Figure 4–1 Hand lines supported by stanchions.

Construction Requirements and Capabilities. Alpine paths can be constructed to almost any desired length. Terrain and weather conditions will affect construction time considerably, but under average conditions 900 meters of path can be constructed by a platoon of engineer troops in 50 to 100 hours, depending on the number of suspended walkways needed. If the men traversing the path are spaced at 4½ meter intervals, approximately 200 men would be on the 900 meter of path at one time. Relative difficulty of the path would dictate their speed of travel. If travel is in one direction only and the path is of average difficulty, 400 men could traverse the 900 meters in 1 hour. Assuming that each man carries a 50-pound load of supplies, 10 tons could be delivered to the terminal end in 1 hour. If the path is constructed so that traffic is free to move steadily in both directions at all points, 400 men or 10 tons of material can be delivered to each terminal every hour.

Figure 4—2 Suspended walkway and stand-off ladders.

4—4. Site Conditions

Route Survey. A reconnaissance or route survey of the terrain over which the alpine path is to be constructed should be made by experienced mountain climbers before construction is started.

Route Requirements. Bypassing of obstacles and attempts at camouflage may require the path to traverse a greater distance but will reduce the construction time, increase the capacity of the completed path, and minimize delay caused by enemy action.

Whenever possible, the path should avoid steep or vertical cliffs, and overhanging rocks. These will require ladders and will consequently slow down personnel using the path.

Snow or rock in areas susceptible to slides must be avoided.

The path must follow a route not visible to the enemy. It is easier to camouflage the route from observation if it follows ravines and rock ledges.

Safety Lines. As a precaution, the personnel performing the route survey should erect safety lines along the route selected as an aid to less experienced men in the construction crew.

- In some locations, safety lines may consist of ropes strung outside the working party, about 1 meter off the ground, and securely fastened to bushes, trees, or temporary stakes.
- In hazardous locations, safety lines may consist of ropes with one end securely anchored. The free end of a rope is tied around the waist of a person engaged in erection. A convenient anchor for such lines can be made from a piton driven into a convenient crack in the rock.

4—5. Material Requirements

Materials. When a route survey has been completed, an estimate of the number and length of each type of component can be drawn up. The materials required for each of the various components are listed in paragraph 4—8 through 4—13. In addition to the equipment listed for the actual construction, erection tools and safety equipment will be required.

Erection Tools. An ample supply of the following tools should be available—

Drill, stone, hand, 4.2 x 60 cm.

Drill, triangular, hand, 2.5 x 45 cm.

Drill, triangular, hand, 1.9 x 45 cm.

Hammer, double-face, striking, 4 pound.

Ladle, hand, 10 cm diameter bowl.

Pot, melting, 18 cm inside diameter.

Pumps, air, hand.

Safety Equipment. A supply of .95 or 1.25 cm diameter fiber rope will be needed for safety lines.

4—6. Transportation

The materials required for the construction of a fixed alpine path can be transported by vehicle or carried by men. When wire or fiber rope is to be carried by men, it may be impractical to cut the rope into sections which make suitable individual loads. In such cases, coil enough rope for an individual load in a figure-eight and leave about 3 meters of free rope before coiling the second load, continuing in the same manner for the succeeding loads. The loads are then carried in a chain fashion.

Section II ERECTION

4—7. General

Size and organization of the working party used for erection will depend on the type of components to be installed, steepness of the terrain, and experience of the troops involved. If the men must carry the materials to the site, allowance must be made for extra men to perform these tasks. A normal working party is a platoon of engineer troops. The men should be organized into groups and assigned to prepare the most difficult sections of the path first. If the men are inexperienced in mountain operations, it will be easier to use a small working party because such a party is easier to control and accidents can be minimized.

4—8. Handlines

General. Handlines are supported by stanchions made of pipe or wooden posts fastened securely in the ground. The ends are fastened to rock anchors to maintain the proper tension in the lines. The lines may be a fiber rope or marline-clad wire rope.

Pipe Stanchions. The standard pipe stanchion is made up of 2.5 cm diameter pipe. Each stanchion is 100 cm long. Two 1.9 wire rope clips (fig. 4—3) are fastened through holes in the pipe with the centers of the clips 90 cm apart. Such a stanchion can be used as is for a suspended walkway which uses 2 wire ropes on each side, but for handlines the lower wire rope clip is removed or left off. Insert the bottom of the stanchion, without the wire rope clip, in a hole drilled at least 25 cm deep.

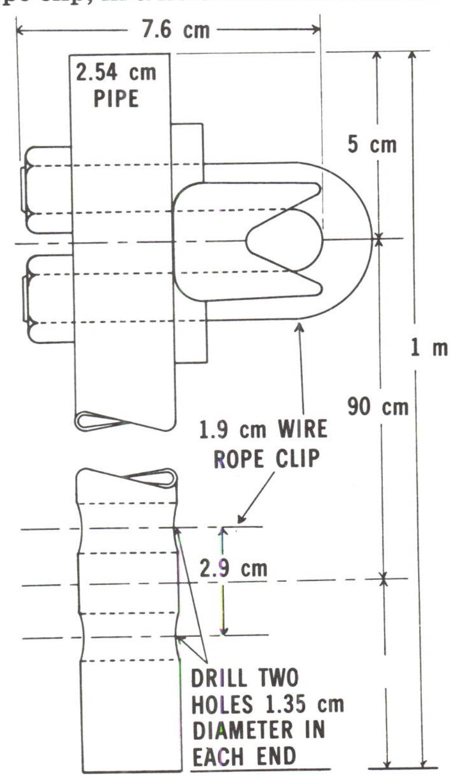

Figure 4—3 Details of an iron pipe stanchion.

Stanchions should be placed in rock wherever possible, using sulfur to hold them in place. Melt the sulfur in a long-handled ladle over a gasoline fire pit and pour it into the hole around the stanchion. If the stanchion must be placed in earth, make the hole in solidly packed ground. Fill the hole around the stanchion with concrete. Space stanchions approximately 10 meters apart on both sides of the path and 75 to 90 cm apart from one side of the path to the other.

Wood Post Stanchions. Wood posts can be used as stanchions (fig. 4—4) for handlines, or standing timber may be used if the path is through natural growth. Wood post stanchions should be 6 to 8 cm in diameter and should be notched near the top so fiber rope handlines can be tied to them with a clove hitch. To tie a clove hitch in the center of a rope, make two turns in the rope close together. Twist them so that the two loops lay back-to-back. Slip these two loops over the end of the stanchion to form the knot.

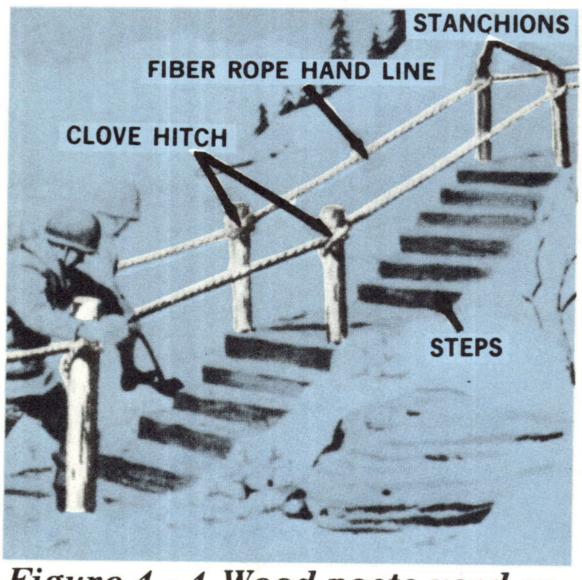

Figure 4—4 Wood posts used as stanchions.

Rock Anchors. Rock anchors are used to hold ends of handlines or other cables. Where handlines are placed parallel to a vertical rock face, they can be supported by rock anchors placed horizontally in the face of the rock instead of by vertical stanchions. Rock anchors used in this manner should be spaced according to the contours of the rock and up to 9 meters apart.

Drilling Holes. There are three general methods of drilling holes in rock for pipe stanchions or rock anchors.

Hand Drilling. In hand drilling, hold the drill steel against the rock with a hammer. After each hammer stroke, rotate the drill slightly so that a round hole will be drilled. One man can hold the drill and stike it with a short-handled hammer. An inexperienced man can drill a hole 25 cms deep in ordinary rock in 30 minutes to 1 hour. Faster work can be done if one man holds the drill and rotates it, while one or two other men strike it with hammers. Three inexperienced men should be able to drill a hole 25 cms deep in about 15 minutes.

Gasoline-operated hammer. A gasoline-operated hammer can be used for drilling holes in positions where there is sufficient room for it to operate, and accessibility permits. It is faster than hand drilling under these circumstances.

Pneumatic drills. Pneumatic drills can be used in more accessible locations and are quite fast. A small portable gasoline-engine-driven compressor is available for such drilling.

Rigging Lines. Wire rope or fiber rope can be used for handlines. However, wire rope is staff and the ends of any broken wires will produce severe cuts on the hands, so that the wire must be served or wrapped before use. A marline-clad rope is available and is a satisfactory material for such uses. Both fiber rope of 1.9 cm diameter and marline-clad wire rope of 1.9 cm outside diameter have the necessary strength. Fiber rope weighs less and is more flexible, but is not as strong as the same size wire rope. Wire rope is more durable, resists weathering, and offers greater resistance to stretching. To rig either rope, pass one end through the eye of a rock anchor at one end of the handline and fasten this free end to the main portion of the rope to provide a firm hold. Fiber rope can be spliced or tied with a bowline, with the eye of the anchor in the bight of the knot. Wire rope can be fastened with wire rope clips or clamps. The coil of rope is now rolled along the path at the base of the stanchions. If wooden post stanchions are used with fiber rope, raise the rope to the top at each stanchion, pull it tight, and tie it around the top of the stanchion with a clove hitch. If pipe stanchions are used, lift the wire or fiber rope to the top of the stanchion and insert a wire rope clip over the rope through the stanchion. Place the nuts on the ends of the U-bolt of the clip and partially tighten them. Pull the rope tight past the stanchion and tighten the nuts securely. Continue along the path in this manner to the end of the handline or until the coil of rope is used up. Fasten the free end to a rock anchor in the same manner as at the beginning of the path. Where rock anchors are used to support the rope along vertical rock faces, the rope is fed through the eyes of the anchors. In rigging wire rope care must be taken to eliminate kinks when they develop along the path, as they will cause stretching and deformation of the wire, weakening the line after erection.

4—9. Steps

Steep and moderate slopes covered with ice or snow make climbing difficult. This is particularly true at high altitudes, where less oxygen is available for breathing. Men tire rapidly under such conditions. Handlines (para. 4—8) are of considerable assistance but do not completely solve the problem. Steps cut in the slope (fig. 4—4) are of material assistance. The riser of the step should not exceed 20 to 25 cm. On gradual slopes a single 20 cm step can be cut every few meters and the surface between steps leveled. On steeper slopes the steps are cut closer together but should not be less than 25 to 30 cm apart. Each step should extend for the full width of the path. For snow and earth use a shovel or pickmattock to cut the steps. Steps in ice are harder to cut and an ice ax, pick, or pickmattock should be used. The steps can be cut before or after handlines are erected, but it is easier to cut them after the handlines are up, since the handlines will be of assistance to the crew cutting the steps.

4—10. Ladders

General. Hanging ladders are made of wire or fiber rope anchored at the top and suspended vertically. They are difficult to ascend or descend, particularly for a man carrying a pack or load, and should be used only when necessary. Stand-off ladders have 2 wood or metal uprights holding them stiff and are placed at an angle which is easier to climb. Both types can be prefabricated to the required size and carried to their final location. At most locations 1 or 2 stand-off ladders are adequate, but hanging ladders act as a traffic bottle neck and 3 or 4 of them should be provided at each location if possible.

Stand-Off Ladders. Stand-off ladders are built in dimensions to suit the circumstances. In general, the two uprights are 5 cm by 10 cm lumber placed on edge, or 8 cm diameter poles roughly cut and dressed at the site. Such uprights are sufficiently strong to extend 5 or 6 meters. Place the uprights 45 cm apart and nail the rungs to them at 25 cm intervals. Use at least 2 nails at each end of each rung. For stand-off ladders made of finished material

Figure 4—5 Construction of stand-off ladders.

(1, fig. 4—5), make the rungs of 2.5 to 5 cm lumber. For ladders made of native material (2, fig. 4—5), make the rungs of 5 cm diameter logs. These logs should be carefully trimmed clean to avoid snagging the clothing of men using the ladder. Anchor each ladder firmly in place with wire or fiber rope fastened about the upper rung and through a rock anchor. The ladder should also be braced at the base, to eliminate all motion during use.

Hanging Ladders. Uprights of hanging ladders may be made of wire or fiber rope and both uprights should be anchored at top and bottom. Wire rope uprights with pipe rungs make the most satisfactory hanging ladders because they are stiffer and do not sag as much as ladders made of other materials. Wire rope uprights with wire rungs can be used. Fiber rope uprights can be used with wood or fiber rope rungs, but should be used sparingly because of their greater flexibility, which causes them to twist under a climber and makes them difficult to use. A rock anchor should be used to anchor each upright of a hanging ladder at the top (fig. 4—6). Place a log at the break of the ladder at the top to hold the uprights and rungs away from the rock face so that better handholds and footholds are provided. A single rock anchor is usually sufficient at the bottom (fig. 4—7) of the ladder, or a pile of rocks can be used as bottom anchor for fiber rope hanging ladders.

Wire rope ladder with pipe rungs. A wire rope ladder can be made using 2.5 or 1.9 cm pipe rungs. The 2.5 cm pipe rungs are most satisfactory. For such ladders the standard pipe stanchion is used. The pipe stanchions (fig. 4—3) are spaced 30 cm apart in the ladder (1, fig. 4—8) and the 1.9 cm wire rope clips are inserted in the stanchion over 1.9 cm wire rope uprights. If .9 cm wire rope uprights are to be used, the wire rope clips are removed from the pipe stanchions and .9 cm wire rope clips inserted in the pipe over the wire rope uprights. When 1.9 cm pipe rungs are used (2, fig. 4—8), the rungs are again spaced 30 cm apart in the ladder but uprights are spaced only 30 cm apart because of the weaker pipe used. The rungs may be fastened in place by two different methods. In one method, a 1.1 cm diameter hole is drilled at each end of each

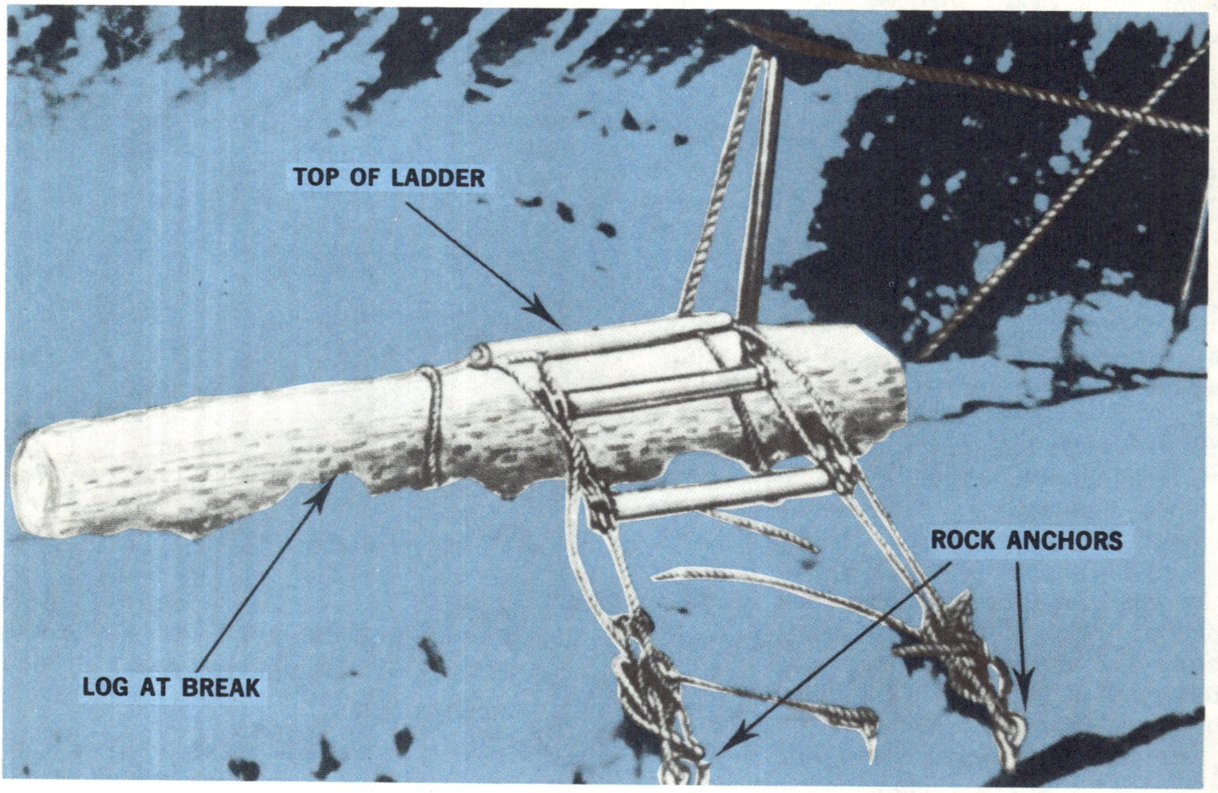

Figure 4—6 ***Method of anchoring top of hanging ladder.***

pipe rung and .9 cm wire rope uprights are threaded through the holes. To hold each rung in place a .9 cm wire clip is fastened about the wire rope upright at each end of each rung, after the rung is in final position. Another method is to cut the pipe rungs 30 cm long and weld the U-bolt of a .9 cm wire rope clip against each end. Space the rungs 30 cm apart on .9 cm wire rope uprights, place the base of the wire rope clips and the nuts on the U-bolts, and tighten the nuts to hold the rungs in place.

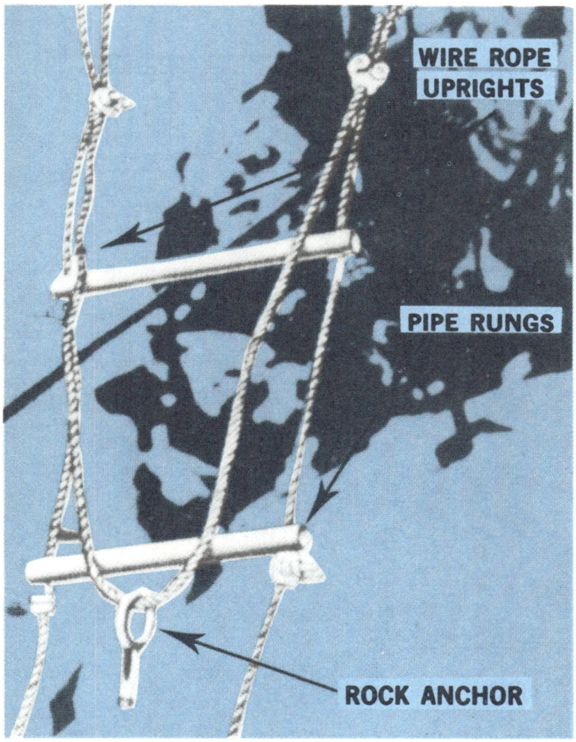

Figure 4—7 Rock anchor method of anchoring bottom of ladder.

Wire rope ladder with wire rope rungs. To make a wire rope ladder with wire rope rungs, lay the .9 cm diameter wire rope uprights out on the ground. Lay the first length out in a series of U-shaped bends. Lay the second length out in a similar manner (3, fig. 4—8) with the U-shaped bends opposing those in the first series and the horizontal rung portions overlapping. Fasten a .9 cm wire rope clip around the overlapping rung portions at each end of each rung to hold them firm.

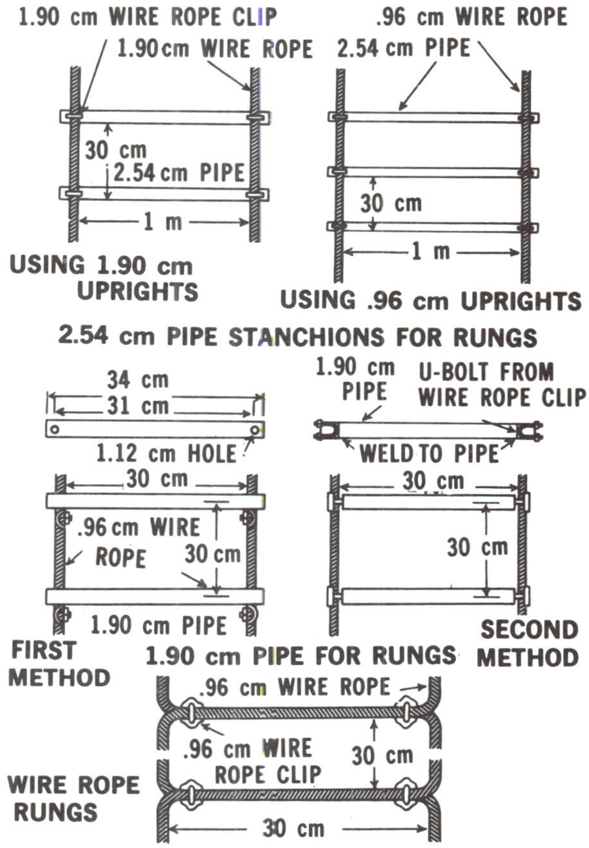

Figure 4—8 Wire rope ladders.

Fiber rope ladder with fiber rope rungs. Fiber rope ladders with fiber rope rungs can be made using 2 or 3 uprights. When 3 uprights are used (1, fig. 4—9), a loop is made in the center upright at the position for each rung. The 2 outside uprights are spaced 50 cm apart. A loop and a single splice hold each end of each rung to the outside upright. A loop in the center of the rung passes through the loop in the center upright. If only 2 uprights are used, the rungs are held in place by-a loop and single splice at each upright. In this case, the 2 uprights must be closer together, with shorter rungs, to stiffen the ladder. Ladders of either type are very flexible and difficult to climb.

Fiber rope ladder with wood rungs. Fiber rope ladders with wood rungs (2, fig. 4—9) can be made using finished lumber or native material. To use native material, cut the rungs of 5 cm diameter material about 40 cm long. Notch the ends of each rung and fasten the rung in the fiber rope upright with a clove hitch. Space the rungs 30 cm apart from center to center. A piece of seizing wire twisted about the back of the clove hitch will make it hold better, but the wire must be placed where it will not snag the clothing of personnel climbing the ladder. If rungs are to be made of finished lumber, cut the rungs to size and drill a 1.9 cm hole at each end. Oak lumber is best for this purpose. Place a .64 cm by 6.3 cm carriage bolt horizontally through each end near the vertical hole to prevent splitting. Tie an overhand knot in the upright to support the rung and thread the upright through the 1.9 cm hole in the rung. Tie a second overhand knot in the upright before threading it through the next rung. Rungs should be spaced 30 cm apart.

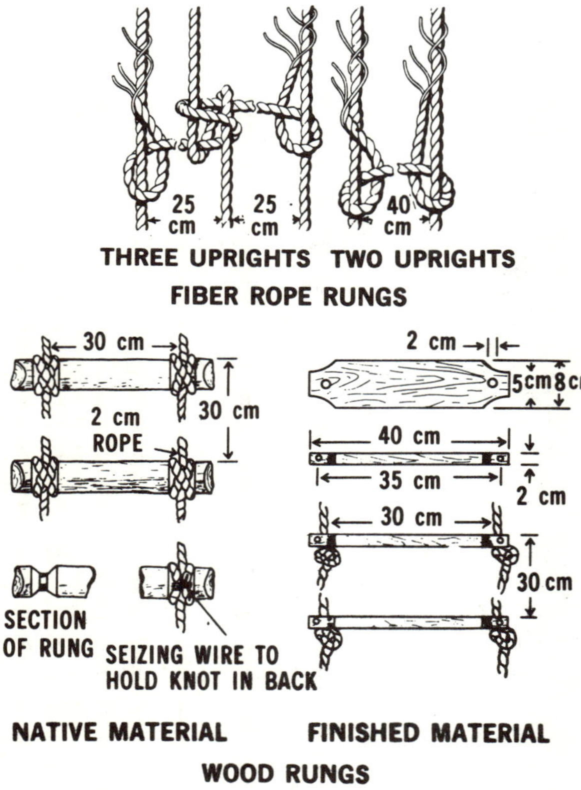

Figure 4—9 Fiber rope ladders.

4—11. Suspended Walkways

Description. A suspended walkway (fig. 4—10) consists of a treadway of either native or finished material suspended like a bridge over a ravine, rocky terrain, or a stream, by 4 wire ropes. The 2 lower ropes support the treadway. Two ropes at a higher level provide handlines, and are reconnected to the lower ropes by vertical stanchions. The wire ropes are rigged and anchored in such a manner that the center of the ropes sag below a straight line between the ends. This sag prevents the tension in the rope from becoming great enough to snap the rope under normal loads and must be computed in advance.

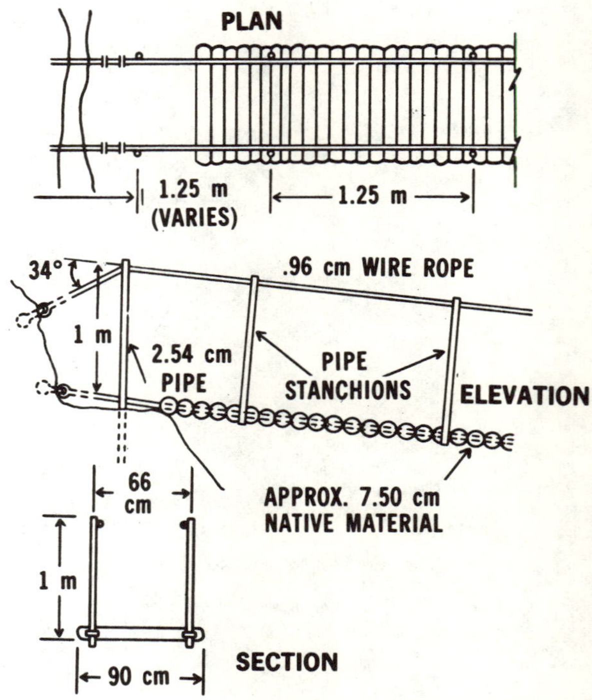

Figure 4—10 Suspended walkway of native material.

Rigging Lines. Four rock anchors and 2 vertical stanchions are required at each end of both terminals before rigging the wire ropes. Fasten each of the lower wire ropes to its anchor at one terminal with wire rope clips; make a turn around the bottom of the end stanchion and pass the end of the rope across the gap to the other terminal. Make a turn in the wire rope around the end stanchion at the new terminal and fasten it to the anchor with wire rope clips. Adjust the wire ropes to the desired sag allowance and tighten the wire rope clips securely at both ends. The upper wire ropes are placed in the same manner, including sag allowance, except that each rope is fastened to its end stanchions with the clips provided on the stanchions.

Anchoring. Drill the holes (para. 4—8) and place the rock anchors. Wherever possible place the rock anchors so that a direct pull is exerted on them, as this develops greatest strength in the anchor. The angle made by the horizontal portion of the upper wire rope (fig. 4—10) and the portion on the rope leaving the anchor should not be greater than 34 degrees.

Sag allowance. The chord span must be measured before the sag allowance is determined. Sag is the vertical distance from the mid point of the cable to the chord (fig. 4—11).

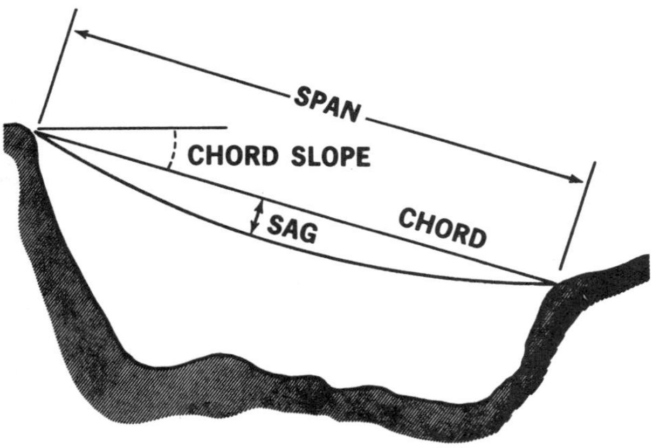

Figure 4—11 Method of computing span and sag of cable.

Sag ratio is the ratio of the sag to the span. For spans up to 12 meters, a sag ratio of 1 percent may be used. For spans between 12 and 20 meters, the correct sag can be determined from the formula:

$$\text{Percent sag} = 1 + 4/3\,\frac{(\text{span} - 12)}{3}$$

With a maximum cable stress of 2,060 pounds for .9 cm wire rope, this formula allows a safety factor of 4, which is the minimum consistent with safety of personnel using the walkway. Where the span of the gap is greater than 20 meters, or the chord slope is more than 20° from the horizontal, a suspended walkway should not be used. Assuming a maximum 20 meter span, the formula above would apply as follows:

$$\text{Percent} = 1 + 4/3\,\frac{(20 - 12)}{3}$$

$$= 1 + 32/9$$

$$= 4.5\%$$

4.5% × 20 meters = .9 meter of vertical sag allowance in cables.

Note: Above computations are approximate and have been rounded off to the nearest whole number.

To install the cable with the sag allowance computed, measure the amount of sag allowance vertically below the wire rope at each end stanchion and mark it clearly. Adjust the wire rope until the line of sight between the two marks on opposite sides of the gap coincides with the lowest point on the wire rope. Sag is allowed in both upper and lower wire ropes in the same manner.

Stanchions. Make up and set the two vertical stanchions at each end of the walkway as outlined in paragraph 4—8. The lower clips are removed or left off these stanchions when they are placed. After the wire ropes are rigged and the correct sag allowance is set, place the remaining stanchions loosely on the wire ropes at one end of the walkway by placing the U-bolts over the wire rope and through the stanchions. Place the nuts on the U-bolts and turn each nut enough to hold it until the treadway is set, but do not tighten it.

Treadway.

Native material. Use logs approximately 9 cm in diameter and 1 meter long. Smooth the logs carefully to prevent injury to personnel using the walkway. The supporting wire ropes are 65 cm apart. Cut a slot about 2.5 cm deep (fig. 4—12) in each end of the logs on the bottom to fit over the supporting wires. Place the logs progressively from one side to the other, pushing the vertical stanchions ahead of the work. As the logs are placed, weave No. 9 galvanized wire, or any similar wire, around them and the wire rope. Pass the first wire over the first log and the second wire under it, then cross the wires and pass the first wire under the second log and the second wire over, as shown in fig. 4—12. Continue weaving the wires in this manner. When the position for the first pair of vertical stanchions is reached, 1 meter from the end stanchion, place an additional vertical notch in the logs to go on each side of the stanchions, as shown in fig. 4—12. At this point, fix one pair of stanchions securely in place by tightening the nuts on the U-bolts firmly. Then place one notched log on the near side of the stanchion, and one on the far side, and weave the wire around them. Continue laying the logs for the treadway as before until the position for the second pair of stanchions is reached. Repeat this procedure as necessary.

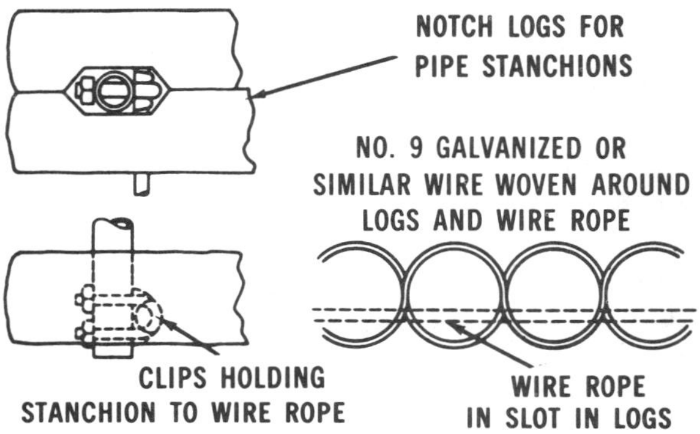

Figure 4—12 Fastening logs in place as treadway.

Native and finished material. Finished lumber 5 cm by 30 cm by 2.5 meters long will make a very satisfactory flooring for a suspended walkway used in combination with native material. Cut and smooth logs approximately 12 cm in diameter and 1 meter long and square them on top for use as stringers (fig. 4—13). Cut a slot in each log at both ends to fit over the wire rope cables. Place one log over the lower wire ropes at a convenient position near one end of the walkway. Place a second log over the lower ropes with its center 1.5 meters from the center of the first log. Saw one 5 by 30 cm board in half and nail one 1.25 meter length in place between the centers of these two logs covering one half the width of the walkway (fig. 4—13). Nail a 2.5 meter length of board in place beside it. Place another log under the free end of the 2.5 meter board and nail it in place. Additional 2.5 meter lengths and logs can now be added to cross the walkway. The joints in the boards will be staggered for greater strength, and an end of board will project each time to provide support for placing each new log stringer. Tie the log stringers at both ends with a wire to hold them firmly to the supporting wire ropes and stanchions.

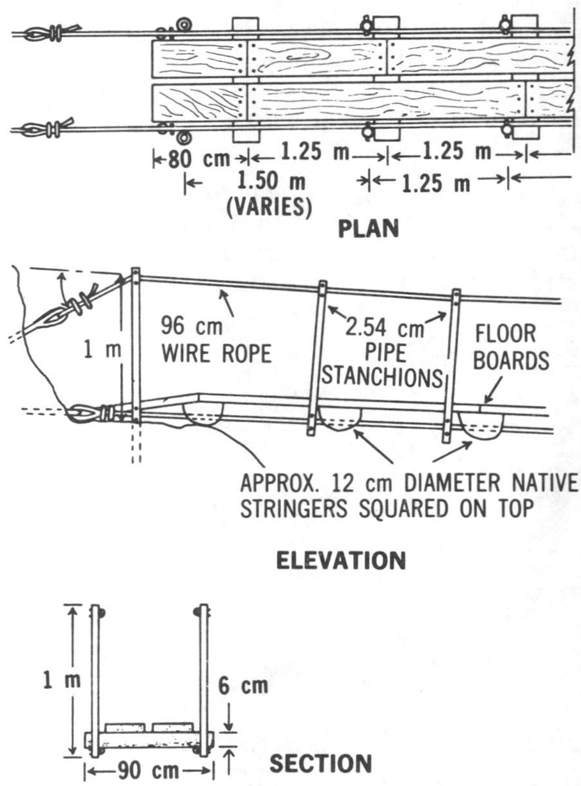

Figure 4—13 Placing treadway of native and finished material.

4—12. Three-Rope Suspended Walkway

A 3-rope suspended walkway is made up of a tread rope and 2 handlines, connected together by stanchions. It is used as an expedient walkway and provides one-way traffic only. Because of the flexibility of fiber rope, wire rope is preferred for these walkways. Considerable motion of the ropes is possible in either case, and personnel using these walkways must proceed with care. When wire ropes are used, the stanchions may be made of wire rope fastened to the main ropes with wire rope clips. Such a walkway is much stiffer and easier to traverse if pipe stanchions are used (para. 4—8) and securely clipped to the main ropes (fig. 4—14). In construction, the 3 main ropes are laid out parallel on level ground and the stanchions are secured to them in pairs at intervals of about 2 meters. The entire assembly is then dragged across the gap and fastened in place. Trees may be used as stanchions or anchors at either end, or pipe stanchions may be set as supports with rock anchors to secure the ends of the lines. One rock anchor is set behind each end stanchion for the handlines, and one is set in the center for the tread rope. Fasten the wire ropes to the anchors securely with wire rope clips at one terminal and adjust the tension to provide the correct sag allowance (para. 4—10) before clipping the ropes to the anchors at the other terminal.

Figure 4—14 Three-rope suspended walkway

4—13. Cableways

Suspended walkways should not be used to bridge gaps greater than 20 meters in length. A cableway should be rigged to span gaps of 20 meters or greater in length. Cableways can be rigged to span gaps as great as 750 meters. Frequently, in constructing an alpine path, some portion of the route requires the erection of one of these cableways. An expedient cableway can be designed and rigged from materials at hand to meet special needs. Detailed construction of cableways is discussed in TM 5—270.

4—14. Inspection

A properly constructed alpine path is capable of withstanding much hard use. However, it must be inspected at frequent intervals to determine the need for repairs. Frequency of inspection should be based on amount of use. If a path is operating at or near capacity, inspection may be required as frequently as twice daily. For lesser use, less frequent inspections are required. Hanging ladders, suspended walkways, and cableways require the most careful inspection. Inspect rope for stretching or signs of wear. Inspect all anchorages for signs of loosened anchors. Inspect wire rope clips for signs of looseness.

4—15. Repairs

Loose wire rope clips are repaired by tightening the nuts. Stretching of wire rope under load may not be objectionable if all lines stretch equally. Unequal stretch should be corrected by loosening the wire rope clips and adjusting the sag in the rope. Provision must be made to prevent the cable from falling before wire rope clips are lossened. Tighten wire rope clips thoroughly after adjustment. Stretched fiber rope in handlines may be tightened at the anchors. Worn steps cut in earth, snow, or ice must be recut when their shape is lost. Lossened anchors can be replaced or moved to new positions. If native material used in ladders or walkways is worn smooth it may be slippery and should be replaced. Frayed or worn fiber rope ladders are an indication of a considerable loss of strength and they should be replaced.

Chapter 5
TRAINING

Section I ORGANIZATION FOR TRAINING

5—1. General

Training is the most important aspect in the conduct of successful mountain operations. A Commander must never disregard or slight the proper training of his command in mountain operations.

A carefully integrated, organized, and staff military mountaineering school is the best means for developing expert military mountaineers. Training should be continuous and uninterrupted for best results. See Appendix B for a suggested training program.

5—2. Selection of Students

Students should be volunteers who understand what they are undertaking.

In the selection of students, steadiness and the ability to improvise are most important; intelligence and physical endurance are only slightly less important.

5—3. Selection of Instructors

In addition to meeting the qualifications of prospective students, instructors should be able to explain clearly and demonstrate precisely both theory and practice when handicapped by lack of uniform instruction areas. Senior instructors should be personnel who have had vast experience as mountain climbers in civilian or military life.

Training:

- Instructors should be given a course closely resembling but in more detail than given to students, plus classes in military methods of instruction.
- Training may be considered complete not only when a candidate can do, but also can explain and demonstrate clearly every operation he has been taught.
- Care should be taken to acquaint instructor classes with varied rock terrain to include all areas in which he will teach.
- Safety precautions must be stressed. Avoidance of accidents depends largely on the instructor's understanding of their causes, and the enforcement of safety precautions.

Duties. The instructor is responsible for the safety and training of his class and the proper care of the equipment. Students should be graded on ability to perform required operations, understanding of subject matter, and special leadership or instructional ability. Continued unsatisfactory grades should result in the rejection of the student receiving them.

5—4. Supervisors

The supervisor oversees the conduct of training. He should:

- Be an experienced and expert mountain climber.
- Make sure that safety precuations are not being overlooked.
- Correct errors of techniques and instruction.
- See that grades and other necessary records are maintained properly.
- Act on all complaints and suggestions personally, if possible, or refer complaints to proper authority.
- Solicit suggestions for improvement of training from instructors and students.
- Insure that classes do not endanger or interfere with each other.
- Encourage interchange of areas by instructors, and encourage regular exchanges of areas with other supervisors.
- Acquaint himself with, and constantly seek to develop, improved techniques and training methods.
- Organize; be prepared to present demonstrations.
- Emphasize all aspects of this training circular to insure that the students and instructors not only learn how to move but also how to fight in mountainous terrain.

5—5. Class Organization

The following organization is suggested:

The company, under the chief instructor is divided into equally sized groups.

Classes are grouped into platoons under the control of the platoon supervisor and his assistant. Size of platoons is governed by the limiting effect of terrain on control. Platoon size should not normally exceed 12 classes. Platoons should be kept intact so far as circumstances permit.

The basic class contains 6 men. However, at the beginning, classes may contain 9 men. Individual soldiers should be grouped according to organization. Officers should be grouped according to component and organization.

5—6. Training Aids

General. Training aids should be used to supplement practical work, but not as a substitute for it.

Training Films. Training films and film strips should be used primarily for orientation. In the absence of appropriate films, conference supplemented by charts are an adequate expedient.

Practice Climbing Tower:

- *Function.* Cliffs suitable for practice of rappel and belay techniques and for safe holding of practice falls may seldom be available. The practice climbing tower will accommodate four classes concurrently.
- *Construction.* (fig. 5—11). The internal frame of the practice climbing tower is constructed of heavy timbers appropriately braced. The front face is constructed of 5 x 30 cm planking. The front face of the tower has a 70° slope.

  The right face of the tower is constructed of 5 x 30 cm planking. This face is also set on a 70° slope.

  The left face of the tower is constructed of vertical peeled native logs 20 to 30 cm in diameter. This face is also set on a 70° slope.

  The rear wall of the tower is vertical and is covered with 5 x 30 cm planking. The top is floored with 5 x 30 cm planks. The safety railing is constructed of 6 cm iron pipe.
- *Utilization.* Four basic classes use the practice climbing tower concurrently, one on each face and one on the rear wall.

  Three ropes may be employed concurrently on each face. The rear wall is normally used for rappelling and for belaying and climbing practice.

  Instructors will normally take their place on top of the tower where they can observe and direct all activities of their particular class. Classes should be rotated from the left or right faces to the rear wall to the front face.

Figure 5—1. *Practice climbing tower.*

338-888 0 - 83 - 8

Section II UNIT TRAINING

5—7. Purpose

The function of unit training is to:

- Familiarize unit commanders with military mountaineering.
- Recondition climbers who have long been absent from mountain terrain.
- Provide necessary training for personnel who are not being trained as expert mountaineers.

5—8. Methods of Training

Orientation. Troops and unit commanders can be oriented by means of a 1-day familiarization course, to include explanation and demonstration of all major phases of training. The demonstration may be spectacular to hold audience interest, but it should not misrepresent either the skills or the proper employment of the climber.

Practical Work. Unit personnel should practice:

- Balance climbing on low pitches with safe landing spots.
- Hauling lines; students should be encouraged to haul each other up and down.
- Handling fixed ropes on a moderately difficult route calculated to increase respect for trained military mountaineers. This, and other phases of practical work, should be under the supervision of instructors and trained unit climbers.

Field Exercises. Having received training in military mountaineering, climbers of units should be given the opportunity to demonstrate and develop their skill in unit problems and field exercises so as to increase their value to their unit.

Conditioning.

- Troops who are not accustomed to operating in mountain terrain or who have not done so for a long time, will lose most of the benefits of specialized training if they are not in strong physical condition. Steep, rocky terrain, gravel pits, quarries, practice climbing towers, and cross-country marches with natural and artificial obstacles, are excellent conditioners.
- Every man must negotiate several rope installations.
- One company in 4 consecutive weeks, training twice a week on the course described below, can acquire the necessary conditioning. The company can be put over the course in 60 minutes. The course should include:
    - A 12-meter suspension traverse between trees, poles, or rocks.
    - A hand-over-hand rope climb ascending a low angle log ramp where feet and legs, and not the hands, are used to lift the body.
    - Descent of low angle slope where a hand-over-hand climb down, or rappel, can be accomplished.
- Properly conducted mountain marches emphasizing proper pace, breathing, and technique of movement are valuable.

Climbing Instructors. The most proficient climbers may be used as instructors in mountain marching and over normally broken terrain. They should be allowed to practice mountaineering under the supervision of the company mountaineering officer.

Section III EXERCISES IN CLIMBING

5—9. General

Purpose. Exercises outlined in this section are only suggestions and may require revision to suit them to existing terrain. They should be short, vigorously executed, and provide climbing action for all participating students. *When action stops, students lose interest.*

Instructors. All instructors should be thoroughly familiar with the purpose and method of the exercise, as well as all climbing techniques and tactics concerned. They should make suggestions necessary for safety or to supplement data omitted from previous training of the students.

Orientation. Orientation of students with all phases of the exercise should precede the problem. Instructors who are charged with the responsibility of moving untrained troops during exercises should be cautioned against the possible reactions when untrained men are exposed to height and rock fall.

Critique. A critique should be held in view of the terrain used in the exercise. As a basis for the critique, errors commonly committed in each exercise should be discussed, and the students should be commended for absence of errors as well as criticized for the errors committed.

5—10. Mountain Walking Course

Purpose. To indicate proper techniques in travel through various terrain types and obstacles.

Method:

- The course should be laid out to include an ascent and descent of about 200 meters in 1 kilometer, passing over or through large and small talus, steep grassy slopes, woods, ledges, easy chimney, down timber, and scree.
- The student travels the course as part of a column of squads under squad control. He is given no coaching ahead of time, but is allowed to use his own judgment.
- Instructors should be posted at various obstacles along the course to observe proper and improper techniques. They should make notes if necessary, and prepare to take part in the critique.

Common Errors and Corrective Action:

- *Talus.* Too much use of hands for balance; poor rhythm in pace, with stops and starts from rock to rock; and stepping too far and too high. The routes should be selected several steps ahead, and the climber should move with a confident rhythmic stride. Silent movement is desirable.
- *Steep grass slopes.* Failure to use uphill side of clumps for steps; selection of too steep a route; and leaning into the hill.
- *Woods.* Tendency on steep slopes to grasp small trees for support, thus shaking the tree crown and disclosing position; branches whipped in faces of others. Trees should be grasped at the base, low enough to eliminate quiver. Branches should be pushed over the head and backward.
- *Ledges.* Tendency to walk toward inside; projections may throw soldiers off balance and where poor footholds are available; tendency to lean in for protection. The climber should choose a route free from cliffs on uphill side, remain upright in posture, taking care not to step on small gravel that might roll underfoot.

- *Easy chimney.* Usually such an obstacle becomes a bottleneck, and delays entire column. Several routes should be taken over this bottleneck to designate assembly point.

- *Down timber.* Entire body raised in stepping over log, loose bark on slippery surface used for foothold. The climber should walk around obstacles if they are at all difficult; step over it rather than upon it, leading with the uphill foot; if stepping upon it, he should not raise the entire body.

- *Scree.* Not enough care in selecting footholds with best support on ascents; leaning back on descents; failure to kick and pack a step properly in traverse to prevent sliding; failure to follow in packed footprints of man ahead.

- *Pace.* Too fast. Proper cadence on long steep marches may vary from 75 steps down to 40 steps per minute and lower. Under heavy load, in high altitude, on 900—1200 meter climbs, or at night, the rest step may be used (short pause after each step).

- *Distance.* Neither great enough nor properly used; excessive accordion action in column, with many starts and stops. With correct distance of 2 meters between men, 10 to 20 meters between squads, 30 to 50 meters between platoons, and with this distance slowly taken up or let out as the terrain requires, no man should have to stop anywhere on the course. Unit leaders must anticipate closing of the gap between units, and shorten and slow his pace to delay or prevent closing of the gap.

- *Route.* Too steep, or failing to make good use of best footing. Route should be no steeper than permits walking with foot flat on the slope rather than on the toes. Lead man should select his footing carefully; others should follow in his footsteps.

5—11. Training of Company Assault Climbers in Attack

Purpose. To indicate correct procedure for the company assault team in attack of a steep rock obstacle.

Method :

- *Obstacle.* A 30 to 60 meter cliff.
- *Situation.* Because of apparent inaccessibility of cliff, top of cliff is lightly defended. No recent enemy action has been observed.
- *Student team.* Six men with basic weapons and appropriate equipment will :
  - Make a visual reconnaissance of the cliff.
  - Give an estimate of the time necessary to establish two fixed ropes up it.
  - Make a plan of attack.
- *Instructor.* Issues fragmentary field order covering situation and execution and prepares a critique with the aid of an observer on top of the cliff.

Common Errors.

- Lack of previous planning; estimate of time, equipment wrong, or lacking.
- Poor reconnaissance; failure of reconnaissance to disclose best routes; noisy reconnaissance.
- Improper placing of fixed rope; operation noisy; pitons poorly placed; rope tied in poorly; poor route selected; security not established at top of cliff; inadequate organization.

The following is a suggested SOP for a company attack:

The assault of the cliff will preferably be made during periods of low visibility. If the night is dark, ropes may have to be set up at dusk or just after dawn. The noise of the movement may be covered by direct and indirect fire. Undefended points should be selected for the attack. The use of at least two separated points increases the likelihood of success.

Before the attack, a complete daylight reconnaissance by the company assault team is desirable. If possible, six routes should be chosen covering a front of from 200 to 400 meters. Of these, the two easiest routes should be picked for the installation of the first two ropes. Concealed locations for mortars are chosen and range cards prepared in advance. Platoon assembly areas are selected and routes to the foot of the cliff decided upon.

Under cover of darkness, the company moves into its assembly area. Mortars go into position to deliver unobserved fire if the attack should be detected. Local security is posted around the assembly area. The best trained men are selected to climb the first rope. They are lined up back from the foot of the cliff and under cover. The assault climbing team (of 6 men each) will initially set up two ropes at points as far apart as possible to give dispersion and still retain control. Numbers 1, 2, 3, and 4 will establish one rope and number 5 and 6 the other.

At the hour of the attack, numbers 1 and 2 climb the cliff anchoring their climbing rope at the top, and immediately move out to provide security. Numbers 3 and 4 follow as closely behind number 2 as possible, place pitons, and secure the first line. Numbers 5 and 6 climb to the top, place their pitons over a different route, and secure their line. They then set up a hauling rope.

Six riflemen selected for their agility follow number 4 up the fixed assault rope and relieve numbers 1 and 2 as security. The riflemen are followed by two observers with radio and wire reel to direct supporting fire. As soon as the hauling rope is set up, automatic weapons are hauled up and put in position to protect the climbing of the main body.

When a rope has been established, one assault climber remains at the rope to service the line and to aid the inexperienced climbers up the route. Other climbers will set up other ropes until four to six are in position, depending upon ease of installation. If the terrain requires, at least two of the routes should have hauling lines and two men at the top at an A-frame to expeditiously assist and remove climbers and loads.

Following observers and automatic weapons, a rifle platoon is moved to the top as rapidly as possible. It reorganizes at the top and takes up a defensive position that will protect the climbers from observed small arms fire. The platoon leader commands all troops at the top of the cliff and integrates direct and indirect fires.

The assault platoon is followed by the company headquarters, artillery liaison party, ammunition, and the remaining riflemen. The 81-mm mortar units and their rifle security are the last troops to climb the rope. The ropes may be left in position for moving supplies or evacuating wounded or may be taken forward to aid in the continuation of the attack.

In using the above standing operating procedure, prior detailed planning and careful organization are essential to success. Additional mountaineering equipment or qualified climbers, if required, are attached from other companies, within the battalion. The initial troops to climb the cliff, including the first platoon, should climb with only their weapons and a minimum of ammunition. The remainder of their equipment will be brought up by hauling lines or carrying parties.

5—12. Equipment Movement Course

Purpose. To indicate possible procedures for the movement of an 81-mm mortar over various types of terrain and obstacles.

Method :

- *Course.* The course should include—

  The ascent of a cliff 10 to 15 meters high, a cliff 20 to 30 meters high, and a steep talus slope over 50 meters long.

  The descent of a steep rock or dirt slope of 40° to 45°.

  The crossing of a gap, such as a steep walled ravine, or between a pinnacle and the main rock face.

- *Student team.* Two men with equipment they have selected after ground reconnaissance, are charged with responsibility of the movement of an

81-mm mortar section over the prescribed course.

- *Instructors.* They are posted at each obstacle. They observe the solution of each problem by the student, and help prepare and assist in delivering a critique.

Common Errors :

- Faulty estimate of equipment needed.
- Best hauling line not used; lines not in best locations.
- Insufficient use of mortar crews to assist in movement of weapons.
- Improperly tied knots; poorly anchored lines; failure to observe common-sense precautions.
- Failure to exercise proper care to avoid damaging weapons.

5—13. Scouts

Purpose. To test the ability of individual climbers, or of two-man climber teams to select a suitable line of march over broken and steep terrain; to recognize those positions of the line of march which will require special aids; to recognize common dangers.

Method :

- *Course.* To include an ascent and descent of at least 300 meters and at least 2½ horizontal kilometers in length; broken and steep terrain, rock cliffs and other terrain obstacles; a wide selection of routes.
- *Student.* Equipped with basic weapon, compass, map, notebook, pencil and climbing equipment, student will proceed from the starting point (indicated on map) toward a designated objective. He will choose the best possible line of march, recognizing the special aids his unit will have to employ in following him. He will make a sketch and make notes of obstacles and danger areas, such as those subject to rock fall from natural and artificial causes. Where terrain bottlenecks occur, he will reconnoiter them and adjacent terrain for alternate routes.
- *Instructor.* He accompanies scout(s) observing and recording mistakes. He assists in preparing and may help to deliver a critique.

Common Errors :

- Failure to make successive visual reconnaissance from commanding terrain features encountered in line of march.
- Poor choice of route; failure to go around, rather than over prominent terrain obstacles; failure to avoid terrain features most subject to enemy observation or swept by rock fall.
- Failure to reach designated point.
- Failure to record obstacles encountered and alternate routes as required.

5—14. Scouting and Patrolling by Night

Purpose. To demonstrate the ability of the patrol to apply the principles of night scouting and patrolling (FM 21—75) to military mountaineering.

Method :

- *Course.* Laid out to include walking on leaf-covered slopes, talus, moderately difficult climbing, easy scrambling, setting up and use of rappel. Difficult climbing should be low (not higher than 3 meters above a good landing spot) and should not exceed a 70° angle.
- *Student.* Single climbers or groups, equipped with a climbing rope, travel course maintaining tactical silence and light discipline.
- *Instructor.* Instructors are posted at frequent intervals along the course. They listen for unnecessary noises and direct patrol from station to station; prepare comments for use in the critique.

Common Errors :

- Rolling or kicking rocks. This error can be corrected by keeping weight over feet, and by moving carefully with special attention to how the rear foot is raised.
- Shaking trees and branches, disclosing position.
- Failure to rappel carefully so that rocks are not knocked loose.
- Failure to avoid skylining.

5—15. Combat Patrol

Purpose. To give the climber team and attached elements practice in functioning as a climber combat patrol.

Practical Work. A tactical problem, in which the climber combat patrol, under control of the company mountaineering officer is called upon to eliminate an enemy combat outpost on dominant and precipitous terrain. For the problem a suitable terrain feature is selected, and a situation and fragmentary attack order are given.

Terrain obstacles are a cliff or spur 15 to 60 meters high, and a talus approach of more than 30 meters.

The situation is tactical. Enemy position is held by a machinegun section and mortar section with rifle security, at a distance of 1,500 meters from friendly forward elements.

The patrol leader is informed of situation and given attack order at H –3 hours. He makes a visual reconnaissance, forms a plan, and briefs his element and climber teams.

The mountaineering noncommissioned officer divides climbers into two units of two men each, and procures necessary climbing equipment, rations, weapons, and ammunition for climbers.

The patrol leader divides his patrol into an assault element and security element—and attaches a climbing team to each. He draws necessary ammunition and rations.

Combat patrol crosses line of departure at H-hour and attack continues according to plan.

Instructors acting as observers accompany the patrol and are stationed with enemy detail on objective. They record their observations and assist in preparing and delivering critique.

Common Errors :

- Failure to organize an attack effectively or procure necessary equipment.
- Assault teams are not coordinated with the security element.
- Climbers do not attack aggressively.
- Climbers warn enemy of their presence prematurely by shouting or dislodging rocks.
- Climbers do not choose best routes; are stopped unnecessarily by difficulty or enemy fire through failure to use covered approach.

REFERENCES

A—1. Army Regulations (AR)

310—25	Dictionary of US Army Terms.
310—50	Authorized Abbreviations and Brevity Codes.

A—2. DA Pamphlets (Pam)

310—1	Index of Administrative Publications, Regulations, Circulars, Pamphlets, Posters, JCS Publications, and General Orders.
310—3	Index of Doctrinal, Training, and Organizational Publications.
310—4	Index of Technical Manuals, Technical Bulletins, Supply Manuals, Supply Bulletins and Lubrication Orders.

A—3. Field Manuals (FM)

5—15	Field Fortifications
5—25	Explosives and Demolitions
5—31	Boobytraps (U)
8—10	Medical Support, Theater of Operations
8—35	Transportation of the Sick and Wounded
21—11	First Aid for Soldiers
21—26	Map Reading
21—60	Visual Signals
21—75	Combat Training of the Individual Soldier and Patrolling
24—18	Field Radio Techniques
24—20	Field Wire and Field Cable Techniques
31—35	Jungle Operations
31—50	Combat in Fortified and Built-Up Areas
31—70	Basic Cold Weather Manual
31—71	Northern Operations
32—5	Signal Security (SIGSEC) (U)
90—6	Mountain Operations

A—4. Technical Manuals (TM)

5—220	Passage of Obstacles Other Than Minefields
5—270	Cableways, Tramways and Suspension Bridges
9—207	Operation and Maintenance of Ordnance Materiel in Cold Weather (0° to —65° F)
10—275	Cold Weather Clothing and Sleeping Equipment

A—5. Common Tables of Allowances (CTA)

50—900	Clothing and Individual Equipment (Active Army, Reserves and DA Civilians)

A—6. Training Films (TF)

FB—78	Avalanche

A—7. Technical Bulletins (TB)

TB Med 288	Medical Problems of Man at High Terrestrial Elevations.

A—8. Training Circulars (TC)

TC 21—3	Soldier's Handbook for Individual Operations and Survival in Cold Weather Areas.

MASTER TRAINING PROGRAM

MILITARY MOUNTAINEERING AND GLACIER WORK

Item	Subject	Code ltr	Total hours	Weeks 1	2	3	4	5	6
1	Rope Management	RM	5	5	—	—	—	—	—
2	Manpacking	MP	1	1	—	—	—	—	—
3	Personal Hygiene and Sanitation	PH	1	1	—	—	—	—	—
4	Use and Care of Climbing Aids	CA	1	1	—	—	—	—	—
5	Preparation of Rations	PR	1	1	—	—	—	—	—
6	Shelters and Fires	SF	5	5	—	—	—	—	—
7	Mountain Weather	MW	1	1	—	—	—	—	—
8	Communications	CM	5	5	—	—	—	—	—
9	Maps and Compass	MC	6	2	4	—	—	—	—
10	Individual Mountain Movement	IM	31	7	8	8	8	—	—
11	Use of Rope Installations	RI	16	4	4	4	4	—	—
12	Assault Climbing Techniques	AC	50	4	12	12	12	10	—
13	Mountain Marches and Bivouacs	MM	54	—	8	16	16	14	—
14	Evacuation	EV	6	2	4	—	—	—	—
15	Glaciers	GL	14	—	—	—	—	14	—
16	Avalanches	AV	2	—	—	—	—	2	—
17	Construction of Alpine Paths	AP	20	—	—	—	—	—	20
18	Commanders' Time	CT	21	5	4	4	4	4	—
	TOTAL	—	240	44	44	44	44	44	20

ROPE MANAGEMENT
Subject Schedule

CODE RM
Total Periods 5

Periods	Type of instruction	Subject	References and remarks
RM 1	Conference and practical work	Types and characteristics of ropes used in mountaineering, care of ropes, proper coiling, tying off of coils, throwing and method of carrying.	TC 90—6—1
RM 2—5	Demonstration and practical work	Terms used in rope work, method of tying knots used in military mountaineering, and their practical application.	TC 90—6—1

MANPACKING
Subject Schedule

CODE MP
Total Periods 1

Periods	Type of instruction	Subject	References and remarks
MP 1	Conference	Nomenclature, function and adjustment of rucksack components, and proper methods of stowing equipment.	FM 31—70 TC 10—8

PERSONAL HYGIENE AND SANITATION
Subject Schedule

CODE PH
Total Periods 1

Periods	Type of instruction	Subject	References and remarks
PH 1	Conference	Prevention and care of ailments peculiar to mountain operations. Personal cleanliness and action to preserve health under adverse conditions.	FM 21—11 FM 31—70

USE AND CARE OF CLIMBING AIDS
Subject Schedule

CODE CA
Total Periods 1

Periods	Type of instruction	Subject	References and remarks
CA 1	Conference	Familiarization with climbing rope, sling rope, pitons, snaplinks, piton hammer, leather gloves, and boots.	TC 90—6—1

PREPARATION OF RATIONS

Subject Schedule

CODE PR
Total Periods 1

Periods	Type of instruction	Subject	References and remarks
PR 1	Conference and practical work	Proper care, use, testing and repair of one burner gasoline stove. Components of field rations and their preparation under field conditions.	FM 21—75 FM 31—70

SHELTERS AND FIRES

Subject Schedule

CODE SF
Total Periods 5

Periods	Type of instruction	Subject	References and remarks
SF 1	Conference and demonstration	Use and care of 5—10 man tents, and the assembly and operation of the Yukon Stove.	FM 31—70
SF 2—5	Practical work	Construction and proper use of various types of improvised shelters. Proper care and use of axe, knife, and the conservation of forest resources. Food procurement and care of the individual under unusual situations.	TC 21—3

MOUNTAIN WEATHER
Subject Schedule

CODE MW
Total Periods 1

Periods	Type of instruction	Subject	References and remarks
MW 1	Conference	The importance of weather in mountain operations, its general characteristics, types and predicition methods.	TC 90—6—1

COMMUNICATIONS
Subject Schedule

CODE CM
Total Periods 5

Periods	Type of instruction	Subject	References and remarks
CM 1	Conference	Wire, radio, messenger, and visual communications, and problems peculiar to their use in mountains. Communications security.	FM 24—18 FM 24—20
CM 2—5	Practical work	Visual Communications.	FM 21—60* FM 32—5*

*5 additional hours to be given each succeeding week as integrated training.

MAPS AND COMPASS
Subject Schedule

CODE MC
Total Periods 6

Periods	Type of instruction	Subject	References and remarks
MC 1—6	Conference, demonstration, and practical work	Review of basic map and compass reading and its application to the special problems caused by difficult terrain.	FM 21—26

INDIVIDUAL MOUNTAIN MOVEMENT
Subject Schedule

CODE IM
Total Periods 31

Periods	Type of instruction	Subject	References and remarks
IM 1—4	Demonstration and practical work	Proper techniques of placing the feet and moving the body while ascending, descending, and traversing varied mountain terrain.	TC 90—6—1
IM 5—12	Demonstration and practical work	Proper technique of using hands, feet, and body position while moving over varied mountain terrain to include rock faces and slab. Margin of safety stressed and students kept close to ground.	
IM 13—20	Demonstration and practical work	Training in the proper technique of descent with aid of rope using all types of rappels.	
IM 21—28	Demonstration and practical work.	Selection of best routes through mountainous areas under all conditions, and proper marking of route.	
IM 29—31	Practical work	Techniques of mountain walking, emplacement of pitons, unroped balance climbing, two-men party climbing, and methods of rappelling.	TC 36—72—1

USE OF ROPE INSTALLATIONS
Subject Schedule

CODE RI
Total Periods 16

Periods	Type of instruction	Subject	References and remarks
RI 1—4	Demonstration and practical work	Installation and use of fixed ropes and suspension traverses.	TC 90—6—1.
RI 5—8	Demonstration and practical work	Installation and use of vertical hauling lines.	
RI 9—12	Demonstration and practical work	Installation and use of rope bridges.	
RI 13—16	Practical work	Movement over a course which requires the use of rope installations.	

ASSAULT CLIMBING TECHNIQUES
Subject Schedule

CODE AC
Total Periods 50

Periods	Type of instruction	Subject	References and remarks
AC 1—4	Conference and demonstration	Demonstration of proper techniques in walking over various mountain terrain to include party climbing, belaying, rappelling, and use of fixed ropes.	TC 90—6—1
AC 5—8	Practical work	Proper techniques of placing the feet and moving the body while ascending, descending and traversing scree, talus, hard ground, and grassy slopes.	
AC 17—24	Demonstration and practical work	Techniques for providing necessary safety factors by proper use of rope. Body, pitons, and natural anchor belay points for roped party climbing.	TC 90—6—1
AC 25—32	Demonstration and practical work	Selection of best routes through mountain terrain and proper marking of route under all conditions.	
AC 33—50	Practical work	Proper techniques of balance climbing and belaying on steep mountainous areas by parties of two or three climbers tied into, and protected by, climbing rope. Adaptation of techniques to overcome lack of various items of equipment.	

MOUNTAIN MARCHES AND BIVOUACS
Subject Schedule

CODE MM
Total Periods 54

Periods	Type of instruction	Subject	References and remarks
MM 1—8	Practical work	Marches through difficult terrain which necessitates application of mountaineering techniques and use of map and compass.	TC 90—6—1
MM 9—54	Practical work	Marches through difficult terrain which requires application of mountaineering techniques and construction and use of improvised shelters. Bivouac routine.	

EVACUATION
Subject Schedule

CODE EV
Total Periods 6

Periods	Type of instruction	Subject	References and remarks
EV 1—2	Conference and demonstration.	Evacuation of casualties over rough terrain cliffs, steep slopes, and across streams. Improvisation of litters and equipment. Use of rope installations as aid in movement.	TC 90—6—1
EV 3—6	Practical work	do	

398-868 0 - 83 - 9

GLACIERS
Subject Schedule

CODE GL
Total Periods 14

Periods	Type of instruction	Subject	References and remarks
GL 1	Conference	Dangers of glacial regions. Mechanics of glacial ice formation. Effects of snow and ice on mountain operations under summer conditions.	TC 90—6—1
GL 2	Conference and demonstration	Special equipment used for movements on snow and ice in high mountains (climbing rope, ice ax, crampons, and ice pitons).	TC 90—6—1
GL 3—10	Demonstration and practical work	Proper techniques of placing the feet and moving the body while ascending, descending, and traversing ice and snow surfaces. Use of ice ax and crampons. Glissading (standing and sitting). Use of rope for party movement. Method of self arrest. Route selection on glaciers.	
GL 11—14	Demonstration and practical work	Proper techniques of crossing crevasses. Crevasse rescue.	

AVALANCHES
Subject Schedule

CODE AV
Total Periods 2

Periods	Type of instruction	Subject	References and remarks
AV 1	Conference	Basic causes of avalanches; types of avalanches; avalanche "triggers."	TF FB—78, Avalanche, running time 30 min. TC 90—6—1
AV 2	Conference and demonstration.	Protective measures (avalanche hazard, forecasting, restrictions, stabilization, and barriers); safety precautions, (observation, route selection, timing, and self defense); techniques of avalanche rescue (equipment, immediate action, followup, and evacuation).	

CONSTRUCTION OF APLINE PATHS

Subject Schedule

CODE AP
Total Periods 20

Periods	Type of instruction	Subject	References and remarks
AP 1—4	Demonstration and practical work.	Cutting of steps in ice and rock. Installation of hanging ladders.	TC 90—6—1
AP 5—8	Demonstration and practical work.	Installation of stand-off ladders.	
AP 9—12	Demonstration and practical work.	Installation of suspended walkways.	
AP 13—16	Demonstration and practical work	Installation and use of hand-operated cableways.	TM 5—270
AP 16—20	Practical work	Movement over a course which requires the use of all items constructed on an alpine path.	TC 90—6—1

COMMANDER'S TIME

Subject Schedule

CODE CT
Total Periods 2

Periods	Type of instruction	Subject	References and remarks
CT 1—21		Reserved for unit Commanders. Mandatory subjects such as character guidance, etc.	5 hour period 1st week, 4 hour period weekly thereafter.

TACTICAL EMPLOYMENT OF CLIMBERS

Requirements for Qualified Climbers

Normally, the first units to be committed to operations in mountainous terrain will be those whose personnel have received this specialized training and have qualified mountaineers assigned or attached as advisors. Depending upon the availability of qualified personnel, a number of *experienced* climbers should be assigned to each platoon, company, or battalion sized unit prior to the planned operations.

Capabilities and Duties of Climbers

To properly perform their duties climbers must be familiar with, and capable of—

- Assistance in unit movements in the mountains such as route selection, navigation, and techniques applicable to the rugged terrain.
- Balance and party climbing.
- Rope installations such as fixed ropes, use of anchors, vertical hauling lines, suspension traverse, and rope bridges.
- Crossing techniques of mountain streams with swift currents.
- First aid and evacuation techniques.
- Individual cooking techniques above and below timberline.
- Shelters and bivouacs.
- Snow characteristics and safety precautions from avalanches.
- Assistance in movements over glaciers.
- Characteristics of weather, especially protection from the sudden changes in temperature, high winds, blizzards, thunderstorms, and other adverse effects.

Military climbers with training in military mountaineering and considerable experience should be used as *guides* and *advisors* to the units involved in mountain operations. They can also be of great value by serving as *observers* and *scouts,* performing duties on patrols of various types, and assisting units in assault climbing.

Guides

Guides are primarily used to assist troops over minor rock obstacles encountered in normally broken terrain. They should possess excellent route-finding and pace-setting abilities and a thorough knowledge of the general practice and theory of mountaineering.

As *advisors* they should—

- Suggest best routes of march.
- Inform unit commanders of natural dangers and obstacles likely to be encountered.
- Be prepared to estimate the time and means necessary for the movement of troops or equipment between any two given points.

In *assisting* troops they should—

- Maintain a steady pace at the head of the column to prevent bunching and straggling.
- Help maintain proper march discipline with respect to pace, cadence, and distance.
- Be stationed at obstacles along the line of march directing troops and indicating alternate routes over bottlenecks.

Observers and Scouts

The rules for the selection of scouts and observers (FM 21—75) hold true in the mountains. To the characteristics and abilities normally required of a scout, agility and thorough training in military mountaineering must be added. The scout should also have a thorough grasp of mountain terrain and terminology, so that he can make accurate sketches and reports of his observations.

Scouts will ordinarily function in pairs. When a scouting mission requires crossing of precipitous terrain, appropriate climbing equipment will be carried.

Reconnaissance Patrols

Reconnaissance patrolling over precipitous rock terrain may best be accomplished by small patrols of one or two pairs of trained climbers.

Flank Patrols

Climbers may expedite movement of flank security elements over difficult rock terrain. One or two climbers should be attached to each element, depending on the size of flank elements and the difficulty of movement.

When ruggedness of terrain is likely to enforce a rate of march on the flanks that is slower than that of the main body, flank security personnel should be grouped near the advance party. Smaller elements are then detached as necessary to reconnoiter and hold dominant terrain features on the flanks of the line of march. Flank outposts join the main body and proceed forward to rejoin the pool of flank elements.

Company Assault Climbers in Attack

One of the principal missions of military mountaineers will be to assist in the attack of very steep objectives. Generally, the forces required to secure this objective from counterattack will be small.

Combat Patrols

The climber team should be the nucleus of patrols operating over steep mountainous terrain. Nonclimbers may be added in varying proportion according to the estimated climbing difficulty and mission.

If more than one point of attack is to be used or a more powerful patrol is needed, additional climbing teams can be employed.

The patrol leader, in addition to being a trained climber, should be familiar with abilities and limitations of each man in the company.

Nonclimbers should be selected with careful consideration of their natural agility and endurance.

MEASUREMENT CONVERSION FACTORS

Multiply	*By*	*To Obtain*
Millimeters	.03937	Inches
Centimeters	.3937	Inches
Centimeters	.03281	Feet
Meters	39.37	Inches
Meters	3.281	Feet
Meters	1.0936	Yards
Kilometers	.62137	Miles
Knots	1.1516	MPH

Ahkio—Boat-like sled used for pulling squad equipment.

Alpine—Of or pertaining to any high lofty mountain generally above the limits of forest growth.

Avalanche—A falling mass of snow and ice.

Avalanche Cord—A brightly colored cord, 15 to 20 meters long, which is tied to the body of a person who is operating in avalanche areas. The cord is a safety measure which facilitates locating and rescuing a person buried by an avalanche.

Bedrock—Unbroken solid rock, overlaid in most places by soil or rock fragments.

Belay—To tie oneself, as a stationary member of a roped party, to a firm rock projection or to a piton, either with the climbing rope or with a snaplink and a sling rope, in order to secure oneself and to afford a safeguard for a moving climber. Belaying is the primary safeguard in mountaineering.

Blizzard—A severe and prolonged snowstorm which may be accompanied by intensely cold high winds.

Chimney—A vertical fissure or wide-crack in a rock or ice face which will admit the body. Many types of chimneys are encountered, from clean splits in the rock or ice varying from 30 to 300 cm in width, to right-angled corners containing cracks.

Company Assault Climbers—Highly trained assault climbing teams who assist in the attack of very steep objectives.

Cornice—An overhanging formation of snow, usually on the lee side of a ridge or at the top of a gully.

Crampon—Metal frame with spikes, fitting the soles of the boots, for use in climbing on hard snow or ice.

Crevasse—A deep fissure, or crack in thick ice or glaciers.

Fissure—A narrow opening in rock, snow, or ice, produced by splitting. A narrow crack.

Firn—Compact, granular snow that is composed of closely associated grains instead of individual flakes. Heat-softened firn is referred to by skiers as spring or corn snow.

Glacier—A river of ice, a few hundred meters to many kilometers in length. The surface of a glacier is broken by crevasses and ice-falls.

Glaze—A coating of ice which forms on exposed objects by the freezing of a film of supercooled water. It occurs under special storm conditions when light rain or drizzle falls through air below 32° F., and strikes a surface that also is below 32° F.

Hummock—A knoll or a small hill. A ridge in an ice field.

Ice Ax—A tool for cutting and scraping steps in hard snow and ice. Also used for a brake and a rudder in glissading and for arresting falls, for steadying the body in climbing up, down, and traversing, steep slopes; as a walking stick, and for probing for crevasses. It is also used to establish a belay.

Knoll—A mound or small hill.

Moraine—Piles of debris consisting of stones, earth and rubble which is carried down by a glacier. There are three types; terminal moraine found at the snout of the glacier; lateral moraine found at the sides; and medial moraine found in the center of the glacier.

Neve—Advanced firn consisting of closely associated grains of ice, separated by air space.

Outcrop—A small cliff.

Piton—An iron peg. The peg has a blade made in various thicknesses and widths so that all kinds of thin cracks can be used in climbing. The piton is hammered into a crack to provide a secure point to which a rope may be hooked by means of a snaplink. They are also used for artificial anchor points for rope installations.

Piton Hammer—A lightweight hammer used for driving pitons.

Rappel (Abseil)—A method of descending over steep rocks, snow or ice, by sliding down a fixed single or double rope.

Ravine—A deep cleft in a mountainside or on the floor of a valley. Normally quite narrow with steep sides.

Rock-Fall—Rock-falls occur on all steep slopes, particularly in gullies and chutes. They are caused by other climbers or by the continual erosion of the rock on a mountain-side caused by freezing and thawing and heavy rain; grazing animals; or enemy action.

Scree—Scree consists of small rocks and gravel which have collected below rock ridges and cliffs or on the side of a mountain. The size of the scree varies from sand to pieces the size of a person's fist.

Slab—A relatively smooth portion of rock laying at an angle, usually 30° to 75° from the horizontal.

Snaplink (Carabiner)—An oval-shaped link made of steel or a light alloy with a spring-loaded gate on one side. The US military snaplink will hold a load of 2,000 lbs. with the gate closed.

Snowbridge—A bridge of snow spanning a crevasse.

Talus—Talus is similar to scree except the rock pieces are larger.

Traverse—To cross mountains, or snow and ice slopes, horizontally.

Travois—A primitive vehicle, consisting of two trailing poles, or shafts, for horses or dogs, and bearing a platform or net for the load. The word as used in this manual pertains to the litter travois, whereby two poles are affixed to the folding litter so they can be used as skids behind the litter as an assist in evacuation down steep terrain.

Tussock—A tuft or clump of growing grass.

Volcanic Tuff—Volcanic ashes that have been mixed with water and dried.

Wall—A rock face inclined between 75° and 90° from the horizontal.

Windchill—The combined cooling effect of wind and temperature on the human body.

ILLUSTRATIONS

TC 90-6-1

30 SEPTEMBER 1976

By Order of the Secretary of the Army:

FRED C. WEYAND
General, United States Army
Chief of Staff

Official:

PAUL T. SMITH
Major General, United States Army
The Adjutant General

Distribution:

Active Army, USAR, ARNG: To be distributed in accordance with DA Form 12-11A, Requirements for The Rifle Co, Platoons, and Squads (Qty rqr block no. 78); DA Form 12-11B, Requirements for Basic Cold Weather (Qty rqr block no. 313); Northern Operations (Qty rqr block no. 314); Mountain Operations (Qty rqr block no. 315).

Additional copies can be requisitioned (DA Form 17) from the U.S. Army Adjutant General Publications Center, 2800 Eastern Boulevard, Baltimore, MD 21220.

U.S. GOVERNMENT PRINTING OFFICE : 1983 O - 398-868 : QL 3

MILITARY MOUNTAINEERING

RANK | DUTY POSITION

We need your IDEAS to make military mountaineering training dynamic and realistic. Please fill out this questionnaire and send us your thoughts. If the questions do not address your particular problem or idea, use any format you wish; just make sure to include your address and AUTOVON phone number so that we can follow up on your suggestion or recommendation.

UNIT ADDRESS | *AUTOVON PHONE NUMBER*

NAME

1. ☐ YES ☐ NO — Does the information given in Chapter 1 give you a feel for what you can expect in military mountaineering? Is the part on weather complete enough or do you need more information? If not what do you need?

2. ☐ YES ☐ NO — Are the techniques in Chapter 2 detailed enough for you to understand them? If not what do you need?

3. ☐ YES ☐ NO — Is there enough information given in Chapter 3 to let you conduct effective operations on glaciers and snow covered mountains? If not what do you need?

4. ☐ YES ☐ NO — Can you effectively plan your own training program using Chapter 5 as a guide? Are the hours and subjects in the Master Training Program adequate? If no, what do you recommend?

5. If you would change anything else in TC 90—6—1, what would it be?

DEPARTMENT OF THE ARMY

POSTAGE AND FEES PAID
DEPARTMENT OF THE ARMY
DOD-314

U.S.MAIL

OFFICIAL BUSINESS
PENALTY FOR PRIVATE USE, $300

Commander
US Army Combat Developments Activity (Alaska)
APO Seattle 98749